Structure-Based Ligand Design

edited by Klaus Gubernator and Hans-Joachim Böhm

Methods and Principles in Medicinal Chemistry

Edited by
R. Mannhold
H. Kubinyi
H. Timmerman

Structure-Based Ligand Design

edited by
Klaus Gubernator
Hans-Joachim Böhm

WILEY-VCH

Weinheim · New York · Chichester · Brisbane · Singapore · Toronto

Series Editors:
Prof. Dr. Raimund Mannhold
Biomedical Research Center
Molecular Drug Research Group
Heinrich-Heine-Universität
Universitätsstraße 1
D-40225 Düsseldorf
Germany

Prof. Dr. Hugo Kubinyi
ZHV/W, A 30
BASF AG
D-67056 Ludwigshafen
Germany

Prof. Dr. Hendrik Timmerman
Faculty of Chemistry
Dept. of Pharmacochemistry
Free University of Amsterdam
De Boelelaan 1083
NL-1081 HV Amsterdam
The Netherlands

Volume Editors:
Dr. Klaus Gubernator
CombiChem Inc.
8050 Camino Santa Fe
San Diego, CA 92121
USA

Dr. Hans-Joachim Böhm
Hoffmann-La Roche AG
Pharma Research
4002 Basel
Switzerland

Library of Congress Card No. applied for.

British Library Cataloguing-in-Publication Data: A catalogue record for this book is available from the British Library.

Deutsche Bibliothek Cataloguing-in-Publication Data:
Structure based ligand design / ed. by Klaus Gubernator ; Hans-Joachim Böhm. - Weinheim ; New York ; Chichester ; Brisbane ; Singapore ; Toronto : Wiley-VCH, 1998
    (Methods and principles in medicinal chemistry ; Vol. 6)
    ISBN 3-527-29343-4

Composition: Datascan GmbH, D-67346 Speyer
Printing and Bookbinding: Franz Spiegel Buch GmbH, D-89026 Ulm

Printed in the Federal Republic of Germany.

Preface

Structure-based ligand design is defined as the search for molecules that fit into the binding pocket of a given target and that can form favorable interactions.

During the past few years, structure-based ligand design has gained an increasingly prominent position within medical chemistry. An impressive number of convincing examples have been published which prove the potential of this approach. Two factors underly the increasing importance and effort of structure-based ligand design: on the one hand, new computer programs for conformational analysis, ligand docking, structural alignment and *de novo* design have been developed; and on the other hand, scientific progress in the molecular biology providing the three-dimensional (3D) structure of therapeutically relevant biopolymers.

Despite such unquestionable progress, lead discovery by structure-based ligand design still faces several problems and limitations such as a lack of availability of 3D structures for the important group of membrane-bound receptors, or adequate computational approaches to reflect induced-fit in drug–reactor interactions.

With regard to the prime aim of our series, i.e. to provide practice-oriented information for medicinal chemists, Klaus Gubernator and Hans-Joachim Böhm have subdivided this volume into two parts. The first part comprises a concise introductory chapter, written by the editors, and covering all aspects of the methodological background in this research field. The second part comprises several chapters which summarize some success stories in the field of structure-based ligand design, such as the design of inhibitors of beta-lactamase (by Gubernator et al.), sialidase (by N. Taylor) or HIV-1 reverse transcriptase (by W. Schäfer). Progress in new computational approaches to predict protein–ligand interactions is described by H. J. Böhm.

The editors would like to thank the contributors to this volume for their cooperation. We are sure that scientists entering the field of structure-based ligand design will find in this volume the adequate support for the successful application of lead discovery techniques.

December 1997

Düsseldorf   Raimund Mannhold
Ludwigshafen   Hugo Kubinyi
Amsterdam   Henk Timmerman

Methods and Principles in Medicinal Chemistry

Edited by
R. Mannhold
H. Kubinyi
H. Timmerman

A Personal Foreword

Writing or editing a book about a rapidly evolving area of science is a challenge. Doing this while being involved in real projects every day is even more of a challenge. And changing location and assignment during this exercise (which is true for both editors) adds further hurdles.

We are thus particularly glad to now present a series of real-life stories from practitioners in the field that have been selected to reflect the current status of structure-based design, it's achievements and also it's controversies. These are embedded in review-type methodological chapters and discussions about scope and future perspectives of these approaches.

Our personal summary from writing several chapters and editing others is that structure-based design actually works and it's successes are now well documented. It cannot be applied to every project and the explosion of biostructural information has just started, dramatically broadening the scope of structure-based design in the future. The methodologies exploiting this information are still evolving and there is room for completely new approaches. This makes this field so interesting, in addition to the beauty of the 3D-structure, and makes writing an account or review so rewarding.

We would like to thank the contributors for the nice collaboration that we had. We also would like to acknowledge the continuous high interest of the series editors in this book project and their very valuable comments and suggestions on earlier versions of this work. We would also like to extend our thanks to our employers who fully support these additional activities, and to our families who have suffered from our mental absence for more than one weekend.

Summer 1997

San Diego   K. G.
Basel   H. J. B.

List of Contributors

Peter Angehrn
Hans-Joachim Böhm
Robert L. Charnas
Ingrid Heinze-Krauss
Christian Hubschwerlen
Christian Oefner
Malcolm P. G. Page
Fritz K. Winkler

Pharma Research
Hoffmann-La Roche AG
4070 Basel
Switzerland

Neera Borkakoti

Roche Products Limited
Broadwater Road
Welwyn Garden City
Herts AL7 3AY
UK

Klaus Gubernator

CombiChem, Inc.
9050 Camino Santa Fe
San Diego, CA 92121
USA

Christine Humblet
Elizabeth A. Lunney

Parke-Davis Pharmaceutical Research
Division of Warner-Lambert Company
2800 Plymouth Road
Ann Arbor
Michigan 48105-2430
USA

Wolfgang Schäfer

Chemical Research Department
Boehringer Mannheim GmbH
D-68298 Mannheim
Germany

Neil R. Taylor

Department of Biomolecular Structure
Glaxo Research and Development Ltd.
Medicines Research Centre
Gunnels Wood Road
Herts SG1 2NY
UK

Contents

1 Rational Design of Bioactive Molecules

K. Gubernator and H. J. Boehm

1.1 Introduction

During the past decade, medicinal chemistry has taken tremendous advantage of the fascinating scientific progress in the field of molecular biology. The three-dimensional (3D) structure determination of biopolymers is one such area which has developed extremely rapidly. As an ever-increasing number of 3D structures is now available, in medicinal chemistry the number of research projects in which the structure of the molecular target is known has rapidly increased. Therefore, there is strong interest both in industry and in academia to develop and apply new approaches that exploit such structural information in the drug discovery process. The purpose of the present volume is to summarize the current state of structure-based design and to present some interesting examples. This should convince the reader that structure-based ligand design has become a mature discipline of medicinal chemistry [1, 2].

Structure-based ligand design is based on the observation that drugs bind to clearly defined molecular targets. A strong and selective binding can be obtained from a high structural and chemical complementarity between the macromolecular target and the ligand [3, 4]. Structure-based ligand design may therefore be described as the search for small molecules that fit into the binding site of the target and can form favorable interactions.

1.1.1 From Ligand Design to Drug Discovery

Structure-based design is now actively pursued by sizeable groups in both academic institutions as well as in the research departments of pharmaceutical companies. Very tight collaboration with both synthetic chemists as well as with experimental biologists performing the biological assay has turned out to be a critical success factor. This often results in the dissemination of structural information to all scientists involved in the project and, ideally, the persistent use of that information in the process of inventing new active principles.

Identifying a ligand to a given target is just the first step in the process of discovering a new drug that offers a therapeutic opportunity for the treatment of a disease. A number of further hurdles have to be taken by a compound until it also qualifies as a drug: It needs to be absorbed, transported, stable to metabolization, and distributed to the right compartment. It also has to be non-toxic, free of side effects, and chemically stable in a formulation. Most of these aspects are even more difficult to assess in a rational fashion. The ligand design part of the drug discovery process is therefore just the first step in a multidisciplinary effort that usually requires several rounds of refinement between the identification of an active principle and the selection of a viable drug candidate. Structure-based design can assist this continuous process by pinpointing opportunities for structural modifications that do not interfere with binding, but alter the properties of the molecule, e.g. its solubility, hydrophobicity or ionization state.

1.2 Source of Structural Information

The most important source of biostructural information still is X-ray crystallography. This requires single crystals of the biopolymers which are often difficult to obtain. In recent years, the availability of reasonable quantities of pure material has improved through the use of biotechnology methods. Protein purification procedures have been improved and more efficient chromatographic methods are nowadays routinely used. Advances have also been made in planning constructs with higher probability to crystallize. Cloning techniques allow us to remove membrane anchors, to express proteins domainwise, and to influence solubility by point mutation of hydrophobic surface residues.

NMR spectroscopy plays an important, complementary role, since it uses biopolymers in solution. High-field magnetic instruments have extended the upper size limit to medium-size systems. NMR spectroscopy requires fairly high concentrations so that solubility of the material is a crucial question. NMR has its particular strength in detecting association phenomena using N15- and/or C13-labeled material [5].

Several thousand structures of macromolecules have been solved, many of which are pharmacologically relevant. The number of newly solved structures continues to grow exponentially due to dramatic methodological improvements in the above-mentioned structure determination techniques.

Most structural information has been generated on enzymes. Their size typically varies from 100 amino acids (e.g., phospholipase A2) to over 500 amino acids (e.g., acetylcholine esterase). In most cases, these are single domain monomeric enzymes, though some enzymes are membrane-bound and anchoring occurs through myristoylation or through N- or C-terminal transmembrane helices.

Other non-enzyme classes of proteins include binding-proteins, antibodies, channels, pores, receptors, membrane proteins and elements of the DNA replication mechanism. Unfortunately, no 3D structure of a pharmacologically relevant membrane-bound receptor has been solved to date.

Structures of complexes of macromolecules with small ligands are important for the design process. Due to very powerful difference density determination methods, more and more of these become available; e.g. more than 200 structures of complexes of HIV protease with bound ligand are known. Complexes of covalently or noncovalantly bound ligands are a valuable source of information about the preferential arrangement of functional groups in tightly bound ligand complexes and can thus give guidance to new design projects.

Most of the structural data are available from the Brookhaven Protein Data Bank (PDB) where published structures are deposited for distribution to subscribers and for on-line retrieval (http://www.pdb.bnl.edu) [6]. Currently, about 6000 structures of more than 600 distinctly different proteins plus 500 DNA structures or complexes with proteins are available.

At Rutgers University, a DNA and RNA structure repository is maintained (http://ndbserver.rutgers.edu) [7]. This contains more DNA structures and is differently annotated.

Both databases also contain NMR structures which are mostly deposited as a collection of low-energy structures satisfying the distance constraints derived from the NMR experiment.

Another important source of structural information is the Cambridge crystallographic structure database (CSD) which contains 150 000 small molecule crystal structures (http://www.ccdc.cam.ac.uk) [8]. This database is an invaluable source of very precise and relevant structural information of almost any class of organic and metallo-organic compounds. It is therefore of crucial importance for structure-based design, both for extracting candidate molecules of known 3D structure and for the deduction of structural rules for implementation in force fields and docking procedures [9]. Since the macromolecular structure field and the small molecule field overlap to a certain extent, many of the larger structures in the CSD are highly relevant for ligand design purposes, e.g., longer peptides and peptidomimetics, DNA oligomers and carbohydrates.

1.3 Classes of Therapeutic Targets

There are several important classes of therapeutic targets. Of the top 100 pharmaceutical drugs, 18 bind to seven-transmembrane receptors, 10 to nuclear receptors, 16 to ion channels and the remainder generally inhibit enzymes. Also, by far the most structural information is available on enzymes. A large fraction of these are globular, soluble proteins which are comparatively easy to crystallize. Examples of enzymes with known 3D structure are: Serine proteases (e.g., elastase [10], coagulation factors [11], see Chapter 2), metalloproteases (e.g., collagenase [12]; see Chapter 4; astacin [13]), aspartyl proteases (e.g., renin [14], HIV protease [15]; see Chapter 3), serine esterases (several lipases [16, 17], acetylcholine esterase [18]; see Chapter 2), dihydrofolate reductase [19], thymidylate synthetase [20], sialidase [21] (see Chapter 6); and betalactamases (see Chapter 5).

G protein-coupled receptors (GPCR) [22] represent a very important class of therapeutic target. Many of the neuroreceptors mediating a neurotransmitter signal from the neuron surface to the intracellular signaling cascade belong to this class. In the absence of experimentally determined high-resolution structures, homology model building has been used to generate 3D structures which could then serve to rationalize structure–activity relationships in a known series of ligands and in the design of new molecules.

It should be noted that other important protein targets exist such as ion channels, transport proteins, and components of the immune system. Structural information for some of these targets exists, but these are beyond the scope of this review.

Significant structural information is also available on DNA and protein–DNA interaction [23, 24]. In order for a drug to interfere with the transcription process, it can either bind to a DNA segment directly or interfere with DNA binding regulatory or transcriptional proteins. Through the availability of structures of DNA intercalators, DNA minor or major groove binding molecules, complexes of DNA with a variety of regulatory proteins as well as transcriptional enzymes [25–27], structure-based design at the DNA level is now possible.

The structural basis described so far can be considerably extended by taking into account the enormous amount of DNA sequence information now available, as well as the protein sequence information derived thereof. It becomes clear that similar protein sequences also exhibit similarity in their 3D structure and often share the same mechanism. Powerful methods for detecting sequence similarity have been developed [28]. Even in the absence of any detectable sequence similarity, threading methods can be used to relate a protein sequence to a known 3D structure [29].

Methods are available [30] to construct 3D models of proteins with a sequence that has similarity to a known structure. The reliability of such a model and its usefulness for ligand design is dependent on the degree of similarity.

Further information can be gained by site-directed mutagenesis. Functional changes in response to changes of amino acids thought to be involved in binding or in the catalytic mechanism allow the validatation of structural hypotheses [31–33]

Many proteins consist of a fairly small set of modular elements [34, 35]. A typical example of these elements is the immunoglobulin-like fold which occurs in a large number of proteins unrelated to antibodies. These domains appear to be recombined to form new functional proteins. Examples of such modular proteins are lectins and lymphokines.

1.4 Protein Ligand Interaction

1.4.1 Covalent versus Noncovalent Inhibitors

Low-molecular weight ligands can interact with macromolecular targets through both covalent and noncovalent interactions. In enzyme inhibition, both cases occur. Most protease inhibitors bind noncovalently (see Chapters 2, 3 and 4). Examples of ligands forming covalent bonds with the target protein are trifluoromethyl ketone [36, 37] and benzoxazinone [38] inhibitors of elastase, β-lactam antibiotics such as penicillin and cephalosporin inhibitors of transpeptidases (see Chapter 5), aspirin, and other nonsteroidal antiinflammatory drugs [39]. Noncovalent and reversible covalent binding is characterized by equilibrium thermodynamics. The binding constant is the equilibrium constant of the association of the inhibitor and the target. For irreversible covalent binding, the inhibition is time-dependent, since the chemical reaction between the inhibitor and the target is usually slower than the nonbonded association.

Only in the case of noncovalent and reversible covalent inhibition, is the association constant K_i a well-determined and reproducible quantity. Exceptions occur when both molecules involved in the association are of fairly high molecular weight; in this case the off-rate of the association becomes very slow and the equilibrium is not reached in the observation period. In cases of covalent inhibition, the inhibitory constants (e.g., IC_{50}) are very much dependent on the experimental conditions (concentration of the constituents, incubation time, temperature, salt, acidity).

1.4.2 Nonbonded Interactions in Protein–Ligand Complexes

Important nonbonded interactions are hydrogen bonds, ionic interactions and lipophilic contacts. Usually, several hydrogen bonds are formed between the protein and the ligand [40]. In addition, oppositely charged functional groups of the protein and the ligand are frequently paired. Furthermore, lipophilic groups of the ligand are found in lipophilic pockets formed by side chains of the hydrophobic amino acids.

What are the important properties that allow a ligand to bind tightly and selectively to a protein? Our current understanding of protein–ligand interactions is still far from being sufficient to answer this question fully. The most important prerequisite appears to be a good steric and electronic complementarity between protein and ligand. However, due to desolvation effects this criterion alone is not sufficient to fully describe tight binding of ligands.

1.4.3 Hydrogen Bonds

The effect of hydrogen bonds on the binding affinity has been analyzed for a set of 80 diverse protein–ligand complexes of known 3D structures and binding constant [41]. If the binding constant K_i is plotted against the number of hydrogen bonds formed between proteins and their ligands, no clear correlation is obtained. Although the binding affinities roughly increase by one order of magnitude per hydrogen bond, there is a very broad scatter. Several protein–ligand complexes exhibit strong binding in the nanomolar range with no H-bonds (e.g., retinol binding protein–retinol [42]) or very few (e.g., antibody–steroid [43], antibody–fluorescein [44]). The binding of these ligands arises from nonpolar interactions. Furthermore, some ligands form many hydrogen bonds with the protein but exhibit only very weak binding (e.g., glycogen phosphorylase–glucose [45], hemagglutinin–sialic acid [46], gluthathione S-transferase–gluthathion [47]). Interestingly, these ligands are all characterized by the absence of significant lipophilic groups.

In contrast, our current knowledge indicates that unpaired buried polar groups at the protein–ligand interface are strongly averse to binding. A recent statistical analysis of high-resolution protein structures showed that in proteins less than 2% of the polar atoms are buried without forming a hydrogen bond [48]. There is a strong tendency to saturate polar groups with hydrogen bonds. Therefore, in the ligand design process one has to ensure that the polar functional groups either of the protein or the ligand will find suitable counterparts if they become buried upon ligand binding .

In addition to hydrogen bonding, there are other important polar interactions in protein–ligand complexes such as coordinative bonds with metal ions. For example, hydroxamic acids and thiols bind strongly to the zinc ion in the active site of metalloproteases [49]. This interaction appears to be the most important contribution to binding in many metalloprotease–inhibitor complexes.

1.4.4 The Role of Solvent in Polar Protein–Ligand Interactions

The contribution of a hydrogen bond depends on its microscopic environment in the protein–ligand complex and can vary drastically. Desolvation effects can completely compensate for the intermolecular hydrogen bond so that the net contribution to binding is close to zero. On the other hand, large contributions to the binding affinity seem to arise from hydroxy groups where this group constitutes an essential structural element of a transition state analog in enzyme inhibitors [50]. As demonstrated by Dao-pin et al. for lysozyme mutants, the effect of salt bridges on protein stability is quite different for buried or solvent accessible ones [51].

An instructive example underlining the importance of desolvation effects was reported by Bartlett et al. for the binding of inhibitors **1** (see Fig. 1) containing phosphonamides (X=NH), phosphonates (X=O) and phosphinates (X=CH$_2$) to thermolysin [52]. The PO$_2$-group interacts with the zinc atom and the NH-group forms a hydrogen bond with the backbone carbonyl group of Ala113. The replacement of this NH-group by a CH$_2$ group does not affect the binding affinity. These results can be understood by comparing the number of hydrogen bonds formed between the inhibitor and the protein or the solvent. The NH group of the phosphonamide forms a hydrogen bond with the protein. It also forms a hydrogen bond with a water molecule in solution. Therefore, ligand binding keeps the total number of hydrogen bonds unchanged. On the other hand, the —CH$_2$— group is not able to form a hydrogen bond with thermolysin or in solution. Therefore, the hydrogen bond inventory is also balanced for the CH$_2$ group. In contrast, phosphonates exhibit a 1000-fold reduced binding affinity. This poor binding is a result of the repulsive electrostatic interaction between the oxygen atom of the ligand and the C=O group of Ala113 which both carry a negative partial charge. The oxygen derivative lacks a hydrogen bond which is presumably present in solution. The X-ray structure of the phosphonate revealed that this inhibitor does indeed bind to thermolysin in the same way as the phosphonamide [53].

In contrast, tight binding can be obtained if the total number of hydrogen bonds increases upon ligand binding. We believe that the complex avidin–biotin [54] achieves tight binding partly through a large number of hydrogen bonds that water molecules cannot form to the same extent at the binding site. The X-ray structures of both the complexed and uncomplexed protein have been determined. The binding site of the uncomplexed protein contains four water molecules, according to the X-ray structure. If we assume that this figure underestimates the actual number of water molecules in the binding site because flexible water molecules cannot be detected in the crystal structure and employ a molecular mechanics calculation to fill the binding site with water molecules, we end up with 9–10 water molecules [55]. These water molecules still form fewer hydrogen bonds with avidin than biotin.

One of the keys to a better understanding of protein–ligand interactions is the role of water. All direct protein–ligand interactions compete with interactions to water molecules. Both the ligand and the protein are solvated before complex formation. They lose part of their solvation shell upon binding, a process which involves various enthalpic and entropic contributions [56–59]. Hydrogen bonds are broken and new ones are formed. Furthermore, the solvent structure is reorganized at the protein–ligand interface. Water has the unique ability to form an extensive network of hydrogen bonds. The hydrogen bond between two water molecules in the gas phase is strong: the dimerization enthalpy is roughly −20 kJ/mol^{-1} [60, 61]. Therefore, if the binding of a ligand simply replaces tightly bound water molecules and does not yield additional interactions, its binding affinity will be small. On the other hand, the replacement of loosely bound water molecules contributes to binding because both entropic (release of water molecules into bulk solvent) and enthalpic factors

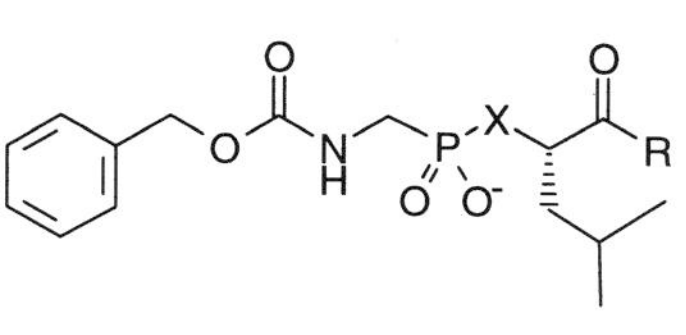

Figure 1. Structure of the thermolysin inhibitor **1**.

(favorable interactions with other water molecules) favor the replacement by the ligand. Unfortunately, detailed investigations of the role of water molecules in protein–ligand interactions have underlined a fairly complex picture [62].

Interestingly, several small molecules containing functional groups frequently involved in protein–ligand interactions do not associate in water. For example, acetic acid does not form a stable complex with guanidine in water [63]. Similarly, simple amides do not associate in water [64]. Apparently, the direct interaction between the small molecules is not strong enough to override the factors favoring the dissociation such as interactions with solvent molecules. As further discussed by Williams and others [56–59], there is also an entropic penalty for the complex formation which may overcompensate favorable interactions.

1.4.5 Lipophilic Interactions

If the binding constant K_i is plotted against the lipophilic contact surface using the same protein–ligand complexes as for the hydrogen-bond statistics, a correlation between the binding affinity and the lipophilic contact surface is observed, but the scatter is again large. An example for ligand binding, completely driven by lipophilic interactions, is the complex of retinol with retinol binding protein [42]. The X-ray structure of this complex shows no hydrogen bonds between the protein and the ligand. The binding constant K_i is 190 nM, corresponding to a free energy of binding of -38 kJ/mol^{-1}. The total molecular surface of retinol amounts to 326 Å^2 [65] of which 295 Å^2 (90%) become buried upon binding. On the other hand, ligands such as SO_4^{2-} have no lipophilic contact surface, but are tightly bound to an appropriate receptor (e.g. SO_4^{2-} and sulfate binding protein with $K_i = 100$ nM [66]). The generally accepted view is that lipophilic interactions are mainly due to the replacement and release of ordered water molecules and are therefore entropy-driven [67, 68]. The entropic gain is due to the fact that the water molecules are no longer positionally confined. There are also enthalpic contributions to lipophilic interactions. Water molecules occupying lipophilic binding sites are unable to form hydrogen bonds with the protein. If they are released they can form hydrogen bonds with other water molecules. In addition, there is a favorable interaction between the lipophilic groups in contact (mainly dispersion interactions).

Recently, the role of specific interactions between aromatic rings has gained increasing attention [69, 70]. A statistical analysis of interaction between phenyl rings in proteins reveals a preference for contact geometries that lead to electrostatically favorable quadrupole–quadrupole interactions [71]. Furthermore, due to their large polarizability, aromatic side chains can also interact with the positive charge of a quaternary ammonium group. This type of interaction is found for example in the complex of acetylcholinesterase with acetylcholine [72].

1.4.6 Criteria for Strong Protein–Ligand Interactions

Our discussion of non-bonded protein–ligand interaction has shown that the quantitative prediction of tightly binding ligands is very difficult. On the other hand, an enormous amount of information has been collected on the structure–activity relationship for ligands binding to proteins of known 3D structure. From the analysis of these data, a number of guidelines have emerged that should always be taken into account when designing a new ligand. They might be summarized as follows:

1. Many tight binding ligands form significant lipophilic interactions with the protein. If the lipophilic contact surface can be enlarged by an additional lipophilic substituent, enhanced binding affinity is frequently observed. Therefore, the search for unoccupied lipophilic pockets should always be one of the first steps in the design of new ligands.
2. Additional hydrogen bonds do not always lead to improved binding affinities but may be nevertheless desirable to improve the selectivity and to make the compound more water soluble. A burial of polar groups upon ligand binding leads to a loss of binding affinity and should be avoided.
3. The binding of a ligand to a protein always leads to the displacement of water molecules. If the ligand can form more hydrogen bonds than the water molecule that are released, then a very tight binding can be achieved.
4. Rigid ligands can bind more strongly than flexible ligands because the entropy loss due to the freezing of internal degrees of freedom is smaller.
5. Water can form strong hydrogen bonds but is not particularly well suited as a transition metal ligand. For transition metal-containing enzymes such as metalloproteases, it is therefore a good idea to incorporate functional groups into the ligands that are known to bind well to metal ions (e.g., thiols, hydroxamic acid).

1.5 Approaches to Structure-Based Ligand Design

In almost any drug discovery project based on a known biochemical target there exists some structural starting point for ligand design. In most of these cases this knowledge is related to the natural substrate (in the case of a processing enzyme), or to a cofactor or the endogenous ligand of the target (in case of a receptor or binding protein). In some cases, such prototype molecules can lead directly to derivatives with drug properties (see the sialidase case). In other cases, structural resemblance to known ligands is unwanted (structurally too complex, too difficult to synthesize, unwanted physicochemical properties, etc.) and other design approaches are favored.

1.5.1 Ligands Derived from Substrate or Natural Ligand

For most of the targets, the natural substrate or ligand is known: thrombin cleaves fibrinogen, HIV protease cleaves the pol polyprotein, the 5HT receptor binds 5-hydroxytryptamine, DD-carboxypeptidases cleave D-alanine from the C-terminus of peptides. Structural analogs of the substrate can be inhibitors if they are no longer processed or transformed by the enzyme. Typically, natural substrates of enzymes can be modified such that the functional group involved in the transformation is no longer viable for the enzyme. If some residual binding affinity exists, the resulting compound is an inhibitor of the respective enzyme; further modifications can improve binding. Prominent examples of this strategy are the aspartyl proteases (Chapter 3) where non-cleavable amide bond mimetics are used as a starting point.

1.5.2 Structures Derived from 3D Database Searches

Substructure searches in topological (2D) databases of compounds of diverse sources (e.g., commercially available compounds, company database) are extensively used by medicinal chemists to preselect substances for screening. Since the pharmacological action is mediated through the 3D shape of the ligand molecule in the receptor-bound conformation, efforts have been made in recent years to construct low-energy 3D structures from 2D databases. A pharmacophore hypothesis can be derived from a series of known inhibitors and their consensus 3D features.

   Two approaches for searching 3D databases have emerged: The first is based on a single conformation of each compound; the search tests the compatibility of each compound with the pharmacophore by generating other possible conformers at the time of the search. In the second, a multitude of representative low-energy conformers are pre-generated (that are likely to contain the receptor-bound conformation) and stored in a multi-conformer database which is then used for searches. The advantage of this latter approach is the enhanced speed of searches; the initial conformation generation step is very time consuming, but has to be performed only once. The program 'Catalyst belongs to the second class. It can be used to generate a pharmacophore hypothesis using a number of reference compounds as input. This hypothesis is then used to search in-house or external structure data bases to find new leads [73].

1.5.3 *De-Novo* Design of Ligands

The sequence and structure information in biostructure research is growing exponentially. Increasingly, we find ourselves in a situation where the structure of an enzyme or a receptor protein is known and suggestions for new inhibitors are desirable. This is where so-called *de novo* design tools are applied [21, 74–76]. In this ambitious approach to structure-based design, it is attempted to construct new molecules completely from scratch. These tools either build up candidate ligands from atoms or fragments, or they search databases of existing structures for complementary molecules. The available programs for *de novo* ligand design have already proven to be an extremely useful addition to the toolbox of medicinal chemistry as they can provide new ideas about possible new ligand structures. A crucial problem of this approach (and also of other approaches to structure-based design) is to control the synthesizability of the proposed molecules. The new field is still in rapid development and is discussed in more detail in Chapter 9.

1.6 Methods and Tools used in Structure-Based Ligand Design

To perform the design of new molecules based on the approaches described above, powerful computer-aided tools are required. These include molecular modeling tools for visualization and analysis, extraction of 3D structures from databases, construction of 3D models using force fields [77–79] and molecular dynamics methods, docking of 3D models to protein cavities. These methods have been documented in detail in the previous volumes of this series and in a number of recent review articles [80–87]. These will therefore only be discussed in the context of the case studies presented in this volume.

1.7 Outlook and Future Developments

During the past few years we have witnessed dramatic improvements in structure-based ligand design. A large number of convincing examples have been published which clearly demonstrate the potential of this approach. Some of them will be described in detail in the following chapters. New computer programs for conformational analysis, ligand docking, structural alignment, and *de novo* ligand design have been described. The first examples for a successful computational design of protein ligands are emerging. The motivation for further developments of these methods is not only the intellectual challenge to understand the nature of protein–ligand interactions, but also the need to speed-up the drug discovery process. As this driving force becomes stronger, the development of advanced methods for structure-based ligand design becomes increasingly important.

Structure-based design considers ligands and their putative interactions with the target. High-affinity binding between a ligand and its target protein is an important prerequisite for the ligand to become a drug. However, focusing on the 3D structure of the target protein carries the danger that other important factors may not be taken properly into account. These are for example transport properties, metabolic stability, oral availability, toxicity, half-life and the addictive potential. They also determine whether a ligand can be applied as a drug. Therefore, physico-chemical properties of the ligand structure should be taken into account at the earliest possible stage.

Despite the remarkable recent progress, the current computational methods for lead discovery still face a number of limitations. Most important is the availability of the 3D structure of the target protein. Substantial efforts are still necessary to develop techniques to make these systems accessible to structure determination. Optimally, their 3D structures are determined together with a series of different ligands. In the absence of such data, it may be possible to construct a model of the target protein from the known structure of a closely related homolog. Otherwise design methods have to rely on comparisons of ligands alone. However, with decreasing level of information about the structural aspects of the protein–ligand system the results from design methods become increasingly less conclusive and reliable.

In cases where the protein structure is available, most current docking and *de novo* design methods treat the protein as rigid. Ligand-induced fit is observed and has to be considered. However, since similar changes of the protein have been observed with several related ligands, the 3D structure of a complex appears to be a good starting point for *de novo* design. The consideration of molecular flexibility of the ligand, especially for species with more than five rotatable bonds, still represents a considerable challenge. This property has to be efficiently incorporated into methods for molecular comparison and ligand docking. Furthermore, the current *de novo* design programs hardly address the problem of the synthetic accessibility of the suggested structures. Finally, current methods for the prediction of binding affinities need to be improved.

Computational methods are only one aspect in the strategy for drug research and development. In the past years, experimental high-throughput screening methods have been established based on molecular test systems. In favorable cases, several thousand compounds can be tested in an assay per day. This has stimulated the development of new methods to synthesize large libraries of diverse molecules. Combinatorial chemistry [88, 89] offers great promise in the discovery of new leads. In our opinion, the computer-based methods nicely complement such screening approaches. Hits from these searches often display rather diverse structures. Molecular comparison methods have to elucidate common features that are likely to be responsible for receptor binding. Once lead structures have been discovered, either through experimental or computer screening methods, molecular modeling and X-ray crystallography come into play to assist efficient lead optimization.

At present, we are beginning to understand some important phenomena and underlying principles determining molecular recognition in protein–ligand complexes. Computational approaches, based on empirical knowledge, provide powerful tools to support the finding and optimization of new lead structures. It can be expected that with the growing experimental data on structures and thermodynamics of protein–ligand complexes the range of applicability of these approaches will be further extended.

The following chapters will cover several aspects of structure-based ligand design. Chapter 2 will give an overview over several projects where structure-based design techniques were successfully employed, while Chapters 3, 4, 5 and 6 each focus on one specific target and describe the use of 3D protein structure in ligand design. It should be noted that structure-based ligand design can also be useful if the 3D structure of the target is not known. This is illustrated in Chapter 7, using the design of inhibitors of reverse transcriptase as an example. In Chapter 8, recent development in *de novo* ligand design are described. Finally, an outlook on the future of structure-based ligand design is given in the final Chapter.

References

[1] Verlinde, L. M. J., and Hol, W. G. J., *Structure* **2**, 577–587 (1994)
[2] Greer, J., Erickson, J. W., Baldwin, J. J., Varney, M. D., *J. Med. Chem.* **37**, 1035–1052 (1994)
[3] Beddell, C. R., (Ed.), *The Design of Drugs to Macromolecular Targets*. J. Wiley, Chichester (1992)
[4] Perutz, M., *Protein Structure: New Approaches to Disease and Therapy*. W. H. Freeman 1992
[5] Billeter, M., *Persp. Drug Discov. Design* **3**, 151–167 (1995)
[6] Stampf, D. R., Felder, C. E., and Sussman, J. L., *Nature* **374**, 572–574 (1995)
[7] Berman, H. M., Olson, W. K., Beveridge, D. L., Westbrook, J., Gelbin, A., Demeny, T., Hsieh, S.-H., Srinivasan, A. R., and Schneider, B., *Biophys. J.* **63**, 751–759 (1992)
[8] Allen, F. A., Davies, J. E., Galloy, J. J., Johnson, O., Kennard, O., Macrae, C. F., Mitchell, E. H., Mitchell, G. F., Smith, J. M., and Watson, D. G., *J. Chem. Inf. Comput. Sci.* **31**, 187–204 (1991)
[9] Klebe, G., *J. Mol. Biol.* **237**, 212–235 (1994)
[10] Powers, J. C., Oleksyszyn, J., Narasimhan, S. L., Kam, C.-M., Radhakrishnan, R., and Meyer Jr., E. F., *Biochemistry* **29**, 3108–3118 (1990)
[11] Banner, D. W., D'Arcy, A., Chene, C., Winkler, F. K., Guha, A., Konigsberg, W. H., Nemerson, Y., and Kirchofer, D., *Nature* **380**, 41–46 (1996)
[12] Borkakoti, N., Winkler, F. K., Willams, D. H., D'Arcy, A., Broadhurst, M. J., Brown, P. A., Johnson, W. H., and Murry, E. J., *Nature Struct. Biol.* **1**, 106–110 (1994)
[13] Bode, W., Gomis-Rueth, F. X., Huber, R., Zwilling, R., and Stoecker, W., *Nature* **358**, 164–167 (1992)
[14] Dealwis, C. G., Frazao, C., Badasso, M., Cooper, J. B., Tickle, I. J., Driessen, H., Blundell, T. L., Murakami, K., Miyazaki, H., Sueiras-Diaz, J., Jones, D. M., and Szelke, M., *J. Mol. Biol.* **236**, 342–360 (1994)
[15] Lam, P. Y. S., Jadhav, P. H., Eyermann, C. J., Hodge, C. N., Ru, Y., Bacheler, L. T., Meek, J. L., Otto, M. J., Rayner, M. M., Wong, Y. N., Chang, C.-H., Weber, P. C., Jackson, D. A., Sharpe, T. R., and Erickson-Viitanen, S., *Science* **263**, 380–384 (1994)
[16] Winkler, F. K., D'Arcy, A., and Hunziker, W., *Nature* **343**, 771–774 (1990)
[17] Cygler, M., Schrag, J. D., Sussman, J. L., Harel, M., Silman, I., Gentry, M. K., and Doctor, B. P., *Protein Sci.* **2**, 366–382 (1993)
[18] Sussman, J. L., Harel, M., Frolow, F., Oefner, Ch., Goldman, A., Toker, L., and Silman, I., *Science* **253**, 872–879 (1991)

[19] Oefner, Ch., D'Arcy, A., and Winkler, F. K., *Eur. J. Biochem.* **174**, 377–385 (1988)
[20] Shoichet, B. K., Stroud, R. M., Santi, D. V., Kuntz, I. D., and Perry, K. M., *Science* **259**, 1445–1450 (1993)
[21] Von Itzstein, M., Wu, W.-Y., Kok, G. B., Pegg, M. S., Dyason, J. C., Jin, B., Van, Phan T., Smythe, M. L., White, H. F., Oliver, S. W., Colman, P., Varghese, J. N., Ryan, D. M., Woods, J. M., Bethell, R. C., Hotham, V. J., Cameron, J. M., and Penn, C. R., *Nature* **363**, 418–423 (1993)
[22] Trumpp-Kallmeyer, S., Hoflack, J., Bruinvels, A., and Hibert, M., *J. Med. Chem.* **35**, 3448–3462 (1992)
[23] Harrison, S. C., *Nature* **353**, 715–719 (1991)
[24] Winkler, F. K., Banner, D. W., Oefner, C., Tsernoglou, D., Brown, R. S., Heathman, S. P., Bryan, R. K., Martin, P. D., Petratos, K., and Wilson, K. S., *EMBO J.* **12**, 1278–1295 (1994)
[25] Werner, M. H., Gronenborn, A. M., and Glore, M. G., *Science* **271**, 778–784 (1996)
[26] Cho, J., Parks, M. E., and Dervan, P. B., *Proc. Natl Acad. Sci. USA* **92**, 10 389–10 392 (1995)
[27] Nunn, C. M., and Neidle, S., *J. Med. Chem.* **38**, 2317–2325 (1995)
[28] Pearson, W. R., *Protein Sci.* **4**, 1145–1160 (1995)
[29] Jones, D. T., Miller, R. T., and Thornton, J. M., *Proteins* **23**, 387–397 (1995)
[30] Johnson, M. S., Overington, J. P., and Blundell, T. L., *J. Mol. Biol.* **231**, 735–745 (1993)
[31] Banner, D. W., D'Arcy, A., Janes, W., Gentz, R., Schoenfeld, H. J., Broger, C., Loetscher, H., and Lesslauer, W., *Cell* **73**, 431–445 (1993)
[32] Dubus, A., Normark, S., Kania, M., and Page, M. G. P., *Biochemistry* **33**, 8577–8581 (1994)
[33] Dubus, A., Normark, S., Kania, M., and Page, M. G. P., *Biochemistry* **34**, 7757–7763 (1995)
[34] Doolittle, R. F., and Bork, P., *Sci. Am.* **269**, 50–56 (1993)
[35] Bork, P., Holm, L., and Sander, C., *J. Mol. Biol.* **242**, 309–320 (1994)
[36] Damewood, J. R., Edwards, P. D., Feeney, S., Gomes, B. C., Steelman, G. B., Tuthill, P. A., Williams, J. C., Warner, P., Woolson, S. A., Wolanin, D. J., and Veale, C. A., *J. Med. Chem.* **37**, 3303–3312 (1994)
[37] Bernstein, P. R., Andisik, D., Bradley, P. K., Bryant, C. B., Ceccarelli, C., Damewood, J. R., Earley, R., Edwards, P. D., Feeney, S., Gomes, B. C., Kosmider, B. J., Steelman, G. B., Thomas, R. M., Vacek, E. P., Veale, C. A., Williams, J. C., Wolanin, D. J., and Woolson, S. A., *J. Med. Chem.* **37**, 3313–3326 (1994)
[38] Spencer, R. W., Copp, L. J., Bonaventura, B., Tam, T. F., Liak, T. J., Billedeau, R. J., and Krantz, A., *Biochem. Biophys. Res. Commun.* **140**, 928–933 (1986)
[39] Silverman, R. B., *The Organic Chemistry of Drug Design and Drug Action*, Academic Press, San Diego, 1992
[40] Jeffrey, G. A., Saenger, W., *Hydrogen Bonding in Biological Structures*, Springer Verlag, Berlin, 1991
[41] Böhm, H. J., and Klebe, G., *Angew. Chem. Int. Ed. Engl.* **35**, 2588–2614 (1996)
[42] Cowan, S. W., Newcomer, M. E., and Jones, T. A., *Proteins* **8**, 44–61 (1990)
[43] Arevalo, J. H., Stura, E. A., Taussig, M. J., and Wilson, I. A., *J. Mol. Biol.* **231**, 103–118 (1993)
[44] Herron, J. N., He, X., Mason, M. L., Voss, E. W., and Edmundson, A. B., *Proteins* **5**, 271–280 (1989)
[45] Watson, K. A., Mitchell, E. P., and Johnson, L. N., *Biochemistry* **33**, 5745–5758 (1994)
[46] Sauter, N. K., Bednarski, M. D., and Wurzburg, B. A., *Biochemistry* **28**, 8388–8396 (1989)
[47] Janes, W., and Schulz, G. E., *J. Biol. Chem.* **256**, 10 443–10 445 (1990)
[48] McDonald, I. K., and Thornton, J. M., *J. Mol. Biol.* **238**, 777–793 (1994)
[49] Powers, J. C., and Harper, J. W., in *Proteinase Inhibitors*, Barrett, A. J., Salveson, G. (Eds.), Elsevier, New York, 244–253 (1986)
[50] Wolfenden, R., and Kati, W. M., *Acc. Chem. Res.* **24**, 209–215 (1991)
[51] Dao-pin, S., Nicholson, H., Baase, W. A., Zhang, X. J., Wozniak, J. A., and Matthews, B. W., *Protein Conformation*, Wiley, Chichester (Ciba Foundation Symposium 161), 199, p. 52
[52] Morgan, B. P., Scholtz, J. M., Ballinger, M. D., Zipkin, I. D., and Bartlett, P. A., *J. Am. Chem. Soc.* **113**, 297–307 (1991)
[53] Tronrud, D. H., Holden, H. M., and Matthews, B. W., *Science* **235**, 571–574 (1987)
[54] Livnah, O., Bayer, E. A., Wilchek, M., and Sussman, J. L., *Proc. Natl Acad. Sci. USA* **90**, 5076–5080 (1993)
[55] Böhm, H. J., unpublished results
[56] Page, M. I., *Angew. Chem. Int. Ed. Engl.* **16**, 449–459 (1977)
[57] Jencks, W. P., *Proc. Natl Acad. Sci. USA* **78**, 4046–4050 (1981)
[58] Searle, M. S., and Williams, D. H., *J. Am. Chem. Soc.* **114**, 10 690–10 697 (1992)
[59] Searle, M. S., Williams, D. H., and Gerhard, U., *J. Am. Chem. Soc.* **114**, 10 697–10 704 (1992)
[60] Chalasinski, G., and Szczesniak, M. M., *Chem. Rev.* **94**, 1723–1765 (1994)

[61] Kim, K. S., Mhin, B. J., Choi, U. S., and Lee, K., *J. Chem. Phys.* **97**, 6649–6662 (1992)
[62] Blokzijl, V. W., and Engberts, J. B. F. N., *Angew. Chem. Int. Ed. Engl.* **105**, 1545–1579 and 1610–1648 (1993)
[63] Spriggs, B., and Haake, P., *Bioorg. Chem.* **6**, 181–190 (1977)
[64] Krikorian, S. E., *J. Phys. Chem.* **86**, 1875–1881 (1982)
[65] Böhm, H. J., *J. Computer Aided Mol. Design* **8**, 243–256 (1994)
[66] Jacobson, B. L., He, J. J., Vermesch, P. S., Lemon, D. D., and Quiocho, F. A., *J. Biol. Chem.* **266**, 5220–5225 (1991)
[67] Tanford, C., *The Hydrophobic Effect*, 2nd Edition, Wiley, New York, 1980
[68] Ben-Naim, A., *Hydrophobic Interactions*, Plenum, New York, 1980
[69] Burley, S. K., and Petsko, G. A., *Science* **229**, 23–28 (1985)
[70] Hunter, C. A., Singh, J., and Thornton, J. M., *J. Mol. Biol.* **218**, 837–846 (1991)
[71] Hunter, C. A., *Chem. Soc. Rev.* **94**, 101–109 (1994)
[72] Sussman, J. L., Harl, M., Frolow, F., Oefner, C., Goldman, A., Toker, L., and Silman, I., *Science* **253**, 872–879 (1991)
[73] Sprague, P., *Persp. Drug Discov. Design* **3**, 1–20 (1995)
[74] Caflish, A., Miranker, A., and Karplus, M., *J. Med. Chem.* **36**, 2142–2167 (1993)
[75] Nishibata, Y., and Itai, A., *J. Med. Chem.* **36**, 2921–2928 (1993)
[76] Müller, K. (Ed.), *De Novo Design*, Escom, Leiden 1995
[77] Müller, K., Ammann, H. J., Doran, D. M., Gerber, P. R., Gubernator, K., and Schrepfer, G., *Bull. Soc. Chim. Belg.* **97**, 655 (1988)
[78] Gerber, P. R., and Müller, K., *J. Computer-Aided Mol. Design* **9**, 251–268 (1995)
[79] Brooks, B. R., Bruccoleri, R. E., Olafson, B. D., States, D. J., Swaminathan, S., and Karplus, M., *J. Comput. Chem.* **4**, 187–217 (1983)
[80] Böhm, H. J., Klebe, G., and Kubinyi, H., *Wirkstoffdesign*, Spektrum Berlin 1996
[81] Weber, H. P. (Ed.), *J. Computer-Aided Mol. Design* **8**, 1–82 (1994)
[82] Colman, P. M., *Curr. Opin. Struct. Biol.* **4**, 868–874 (1994)
[83] Cohen, N. C., Blaney, J. M., Humblet, C., Gund, P., and Barry, D. C., *J. Med. Chem.* **33**, 883–894 (1990)
[84] Böhm, H. J., *Biotech. Curr. Opinion* **7**, 433–436 (1996)
[85] Kuntz, I. D., *Science* **257**, 1078–1082 (1992)
[86] Borman, S., *Chemical & Engineering News*, 10, August 1992, pp. 18–26
[87] Kuyper, L. F., Roth, B., Baccanari, D. P., Ferone, R., Beddell, C. R., Champness, J. N., Stemmers, D. K., Dann, J. G., Norrington, F. E. A., Baker, D. J., and Goodford, P. J., *J. Med. Chem.* **25**, 1120–1122 (1982)
[88] Gordon, E. M., Barrett, R. W., Dower, W. J., Fodor, S. P. A., and Gallop, M. A., *J. Med. Chem.* **37**, 1385–1401 (1994)
[89] Weber, L., Wallbaum, S., Broger, C., and Gubernator, K., *Angew. Chem. Int. Ed. Engl.* **34**, 2280–2282 (1995)

2 Examples of Active Areas of Structure Based-Design

K. Gubernator and H. J. Boehm

Structure-based ligand design can contribute significantly to the drug discovery process at several steps. In the present chapter this is highlighted by several examples. Propably the most important application is the design of completely novel molecules (e.g. thrombin inhibitors, section 2.1). In the next step, structure-based design can be used to modify a known ligand to improve the binding affinity and/or the pharmacokinetic properties (sections 2.2 and 2.3).

Another important area of research in structure-based design is the elucidation of possible binding modes and the rationalization of observed structure–activity relationships. This aspect is highlighted by the examples described in sections 2.4 and 2.5.

2.1 Thrombin Inhibitors

Thrombin, as well as other coagulation factors, is a trypsin-like serine protease. The mechanism of action of this class of enzymes is well understood based on a large number of structural studies [1]. These include crystallographic work with substrates and derivatives, with mechanism-based inhibitors and with noncovalent inhibitors, as well as NMR studies with substrate peptides [2–4]. To summarize, the scheme of the substrate cleavage by trypsin and a model of the transition state of the acylation reaction are shown in Figs. 1 and 2. The transition state is characterized by a tetrahedral carbon covalently linked to the catalytic serine side-chain oxygen atom, a negatively charged oxy-anion stabilized by two hydrogen bonds to two backbone NH groups, and a protonated histidine which has received the proton from the serine-OH.

The three-dimensional structure of the transition state and the protein environment exhibit ideal complementarity to each other, which is the prerequisite for the observed rate enhancement of the hydrolytic reaction.

The specificity for cleavage after arginine or lysine is mediated by a recognition pocket with an aspartate side chain at the bottom. This pocket accommodates the positively charged side chain by forming a salt bridge with the carboxylate of the aspartic acid.

In thrombin, the specificity extends far beyond the residue preceding the cleavage site. NMR [4] and X-ray studies [3, 5] revealed that three hydrophobic residues in fibrinogen **2** two, eight and nine residues before the cleavage site, respectively, occupy unique hydrophobic pockets (Fig. 3). These hydrophobic pockets are formed by a unique additional loop Tyr-Pro-Pro-Trp above the active site.

The binding mode of three previously known inhibitors MD805 **3** and NAPAP **4** (Fig. 4) as revealed in the X-ray structure of their complexes with thrombin [2, 3] has some unexpected

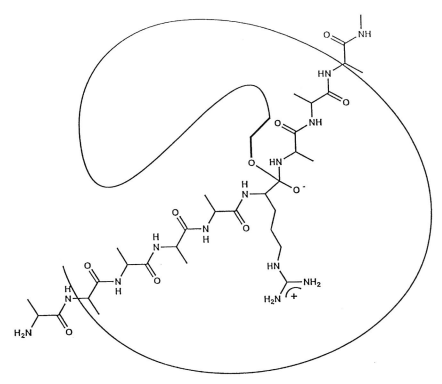

Figure 1. Schematic representation of the cleavage of a peptide **1** by trypsin at the amide bond after an arginine residue.

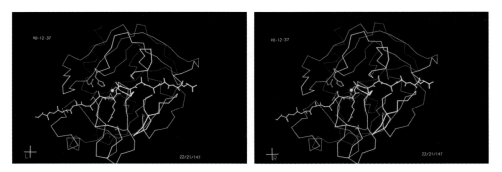

Figure 2. Three-dimensional model of the peptide **1** cleavage by trypsin based on the X-ray structure of the protein.

features: The basic moiety binding in the specificity pocket is directly connected to hydrophobic portions occupying the two hydrophobic pockets; there is no interaction with the catalytic serine. The inhibitors thus do not bind substrate-like, but bypass the catalytic site and

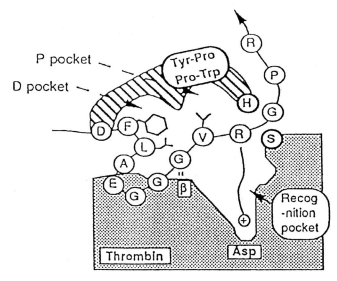

Figure 3. Cartoon of the binding mode of fibrinogen **2** in thrombin with the arginine side chain in the recognition pocket, the valine side chain in the P-pocket and the leucine and phenylalanine side chains in the D-pocket.

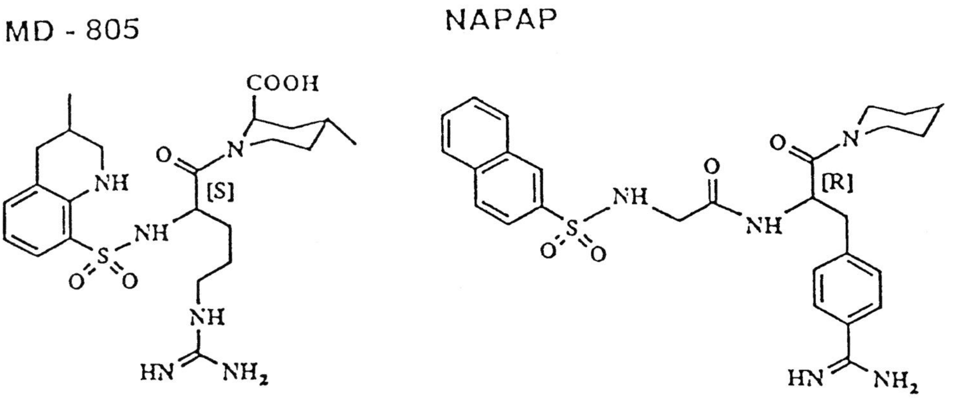

MD - 805   NAPAP

Figure 4. Structure of the MD805 **3** and NAPAP **4** thrombin inhibitors.

preferentially interact with the hydrophobic pockets (Fig. 5). In the complex with fibrinogen, the same pockets are occupied by the hydrophobic side chains preceding the cleavage site.

Based on these findings, two new classes of potent thrombin inhibitors have been discovered by a three-step process [6–8]. First, a collection of small, basic molecules has been tested in thrombin and trypsin assays; *N*-amidinopiperidine has been selected as being more active than benzamidine and slightly selective for thrombin. Secondly, substituents attached to the 3-position of amidinopiperidine containing two hydrophobic moieties mediate low nano-

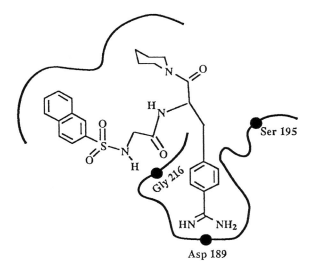

Figure 5. Schematic representation of the experimentally determined binding mode of NAPAP **3** to thrombin.

molar activity [9] (Fig. 6). Replacement of the D-amino acid middle portion by a sidechain-connected L-aspartate finally leads to picomolar thrombin inhibitors which are highly selective (Fig. 7) [10].

Several of these compounds have been studied by X-ray crystallography of complexes with human thrombin; two representative examples are shown in Figs. 6, 8 and 7, 9, respectively.

Interestingly, they exhibit two different binding modes where the occupancy of the two neighboring hydrophobic pockets is distinctly different. In the case of the D-amino acid derivatives, the naphthyl portion occupies the rear side of the pockets toward His57; in the aspartate case, the substituent on the glycine nitrogen occupies the inner pocket and naphthylsulfonyl unit the outer one.

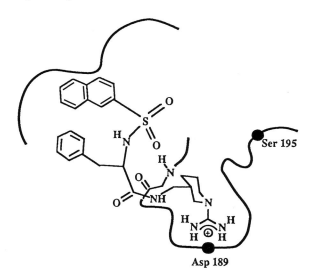

Figure 6. Schematic representation of the experimentally determined binding mode of the D-Phe derivative **5** to thrombin.

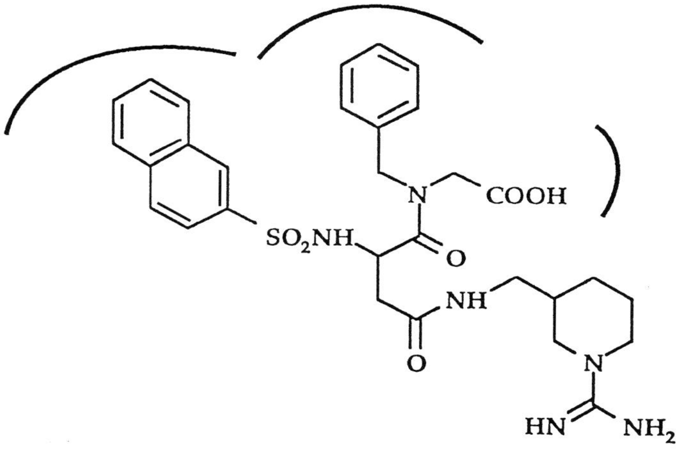

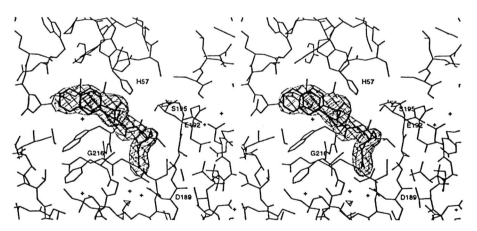

Figure 7. Schematic representation of the experimentally determined binding mode of the aspartate-benzyl-glycine derivative **6** to thrombin.

Figure 8. Difference electron density map for the D-Phe derivative **5**.

From the latter class of compounds, the *N*-cyclopropylglycin derivative **7** (Ro-46-6240) has been selected for clinical development as a possible intravenous antithrombotic drug based on pharmacokinetic considerations (Fig. 10.)

2.2 Design of Orally Active Inhibitors of Elastase

The natural substrate is a good starting point for the design of protease inhibitors. If the scissile bond is replaced by a noncleavable isoster, potent inhibitors are frequently obtained. Alternatively, one may try to introduce functional groups that form a covalent bond with the

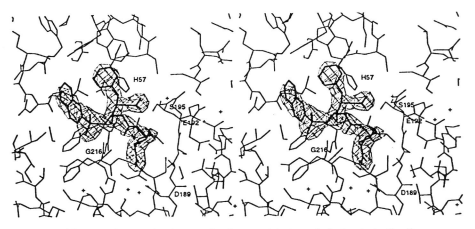

Figure 9. Difference electron density map for the aspartate-benzyl-glycine derivative **6**.

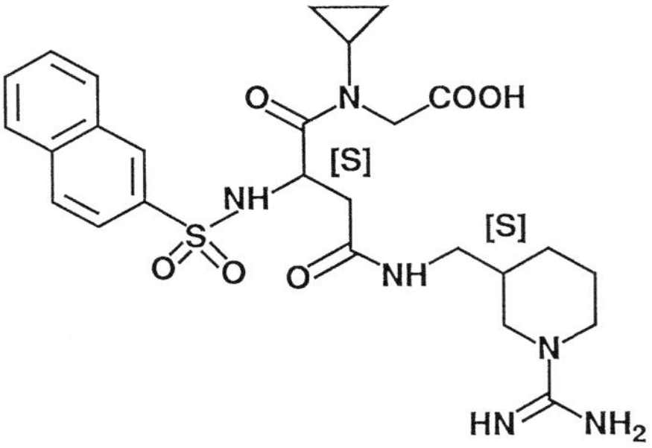

Figure 10. Chemical structures of **7** (Ro-46-6240).

enzyme. The challenge is then to convert the peptidic lead into a nonpeptidic compound that is metabolically stable and orally active. The work from the Zeneca laboratories on elastase inhibitors is a nice example for the design of nonpeptidic compounds based on the 3D-structure of the target protein complexed with a peptidic inhibitor [11].

Human leukocyte elastase (HLE) is a serine protease which is used in the lung to degrade necrotic tissue and invading bacteria. Under normal conditions, the activity of elastase is controlled by several endogenous inhibitors such as α1-protease-inhibitor. If the balance between protease and inhibitor is shifted, then elastase also attacks healthy tissue, leading to emphysema.

A potential approach to treat this life-threatening desease is the use of elastase inhibitors. At Zeneca/ICI, a project was initiated to find new low-molecular weight inhibitors of elastase. First, substrate-analog peptidic inhibitors were investigated. Trifluoromethylketones R—COCF3 are known to bind tightly to serine proteases as covalently binding, reversible inhibitors. Therefore, the design focussed on this class of inhibitors. Starting from the peptidic substrate sequence, very potent elastase inhibitors were discovered, e.g., **8** and **9** (Fig. 11).

In a clinical evaluation, ICI 200880 (**9**) proved to be a very effective elastase inhibitor. However, the compound is not orally active and has a very short duration of action. By that

8

9

ICI 200880

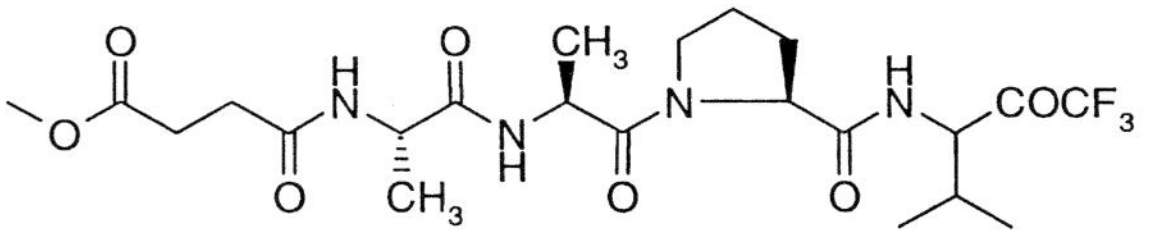

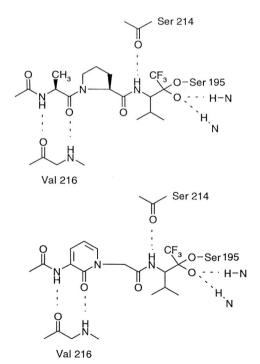

Figure 11. Substrate-analoge inhibitors of elastase **8** and **9**. Compound **9** is highly potent *in vitro* but is not orally active.

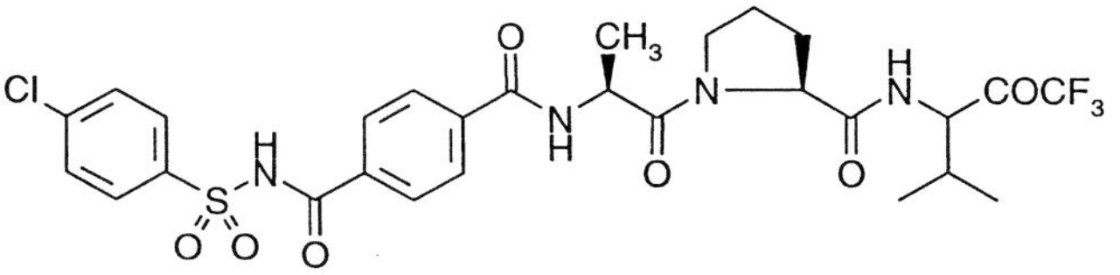

time, the 3D structure of elastase complexed with the closely related inhibitor Ac-Ala-Pro-Val-CF3 was solved. The most important interactions between elastase and the inhibitor are shown in Fig. 12. The inhibitor binds in a β-sheet conformation to elastase, forming two hydrogen bonds to Val216 and one hydrogen bond with Ser214. The valine side chain occupies the P1-pocket and the carbonyl group binds as a hemiacetal to the sidechain of Ser195. The most important information from the X-ray structure is the knowledge of those functional groups

Figure 12. Comparison of the binding mode of the elastase-inhibitor Ac-Ala-Pro-Val-CF3 with the postulated binding mode of the pyridones (e.g., **10**, Fig. 13). Both compounds should be able to form two hydrogen bonds to Val216.

that interact directly with the enzyme and their spatial relationship. The goal was now to find new functional groups that could form the same set of interactions as the peptidic inhibitor.

Starting from the 3D structure of the protein–ligand complex, a number of possible structures were investigated using model building. It was decided to investigate pyridones as possible inhibitors of elastase. According to molecular modeling studies, this type of inhibitor should fit into the binding site and should be able to form the same pattern of hydrogen bonds as the peptide. The postulated binding mode of the pyridones is compared to the binding mode of the peptide in Fig. 12. Several compounds from this class were synthesized. As expected, they are potent inhibitors of elastase. Compound **10** (Fig. 13) binds to elastase with $K_i = 5.6$ nM. Unfortunately, this compound turned out not to be orally active. Furthermore, it is not selective and also binds to chymotrypsin ($K_i = 60$ nM). The poor oral bioavailability was thought to be due to the high lipophilicity ($\log P > 4$) and the low solubility in water. In addition, the synthesis of the pyridones is difficult. It was therefore decided to shift

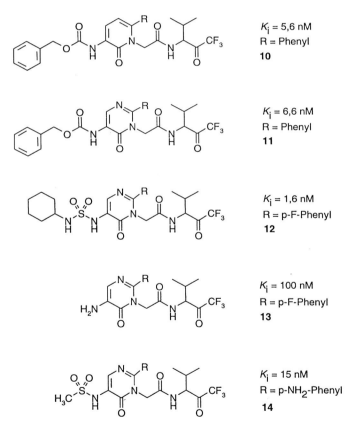

$K_i = 5.6$ nM
R = Phenyl
10

$K_i = 6.6$ nM
R = Phenyl
11

$K_i = 1.6$ nM
R = p-F-Phenyl
12

$K_i = 100$ nM
R = p-F-Phenyl
13

$K_i = 15$ nM
R = p-NH₂-Phenyl
14

Figure 13. Design of orally active inhibitors of elastase at Zeneca. The initial idea, to replace the Ala-Pro unit by a pyridone, led to **10**. Later, pyrimidones were investigated. In this structural class, very potent elastase inhibitors were found, e.g., **12**. Very good *in vivo* properties are observed for **13**. The *p*-fluorophenyl- or *p*-aminophenyl substituent improves the lipophilic contact with the enzyme.

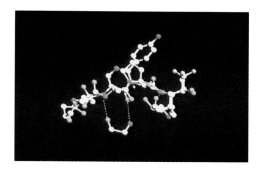

Figure 14. Comparison of the 3D structures of the elastase inhibitor **14** and MeO-Suc-Ala-Pro-Val-CF3, both bound to elastase. The peptidic inhibitor is shown in green. **14** is depicted in a color-coded representation. Both inhibitors form two hydrogen bonds with Val216 and one hydrogen bond with Ser214. In addition, both inhibitor place an oxygen atom into the oxygen anion hole.

the synthetic effort to pyrimidones which can be synthesized more easily. This class of compounds exhibits significantly improved biological characteristics. Compound **11** is less lipophilic ($\log P = 2.1$), ten times more water-soluble and orally active. The binding affinity to elastase remains the same ($K_i = 6.6$ nM) and chymotrypsin-inhibition is strongly reduced ($K_i = 1000$ nM).

A large number of representatives of this new class of elastase inhibitors were synthesized and tested for enzyme inhibition and bioavailability. It emerges that strong inhibition and *in vivo* activity are not correlated. For example, **12** is a highly potent elastase inhibitor, but has no oral activity. The best disclosed compound is **13** ($K_i = 100$ nM), the bioavailability of which in hamster, rat and dog is in the range of 60–90 %. The duration of action is more than 4 hours. Compound **13** (Fig. 13) therefore appears to be an interesting candidate for clinical evaluation. The 3D structure of elastase from porcine pancreas complexed with the inhibitor **14** was determined (Fig. 14). The binding mode of the inhibitor is exactly as predicted by molecular modeling.

In our view, the work described above is a very nice example for structure-based design. The key to success is the use of all available structural information and an early consideration of the physicochemical properties.

The described work has also significant impact on the design of inhibitors for other enzymes. The first published nonpeptidic inhibitors of the cysteine protease interleukin converting enzyme (ICE) have been designed following the Zeneca approach [12]. Similarly, thrombin inhibitors have recently been disclosed that were designed along the same lines [13].

2.3 Dorzolamide: A Success Story of Structure-Based Drug Design

The application of structure-based design has already contributed significantly to the discovery of marketed drugs. We would like emphasize this point by briefly summarizing some structural aspects in the discovery of the carbonic anhydrase inhibitor dorzolamide [14, 15].

The ACE-inhibitor captopril is frequently quoted as the first example of a drug which was obtained from structure-based design. Indeed, in the paper describing the discovery of captopril, Cushman and coworkers emphasize the value of their active site model of ACE in the design of captopril [16]. As the 3D structure of ACE was not known by Cushman, the work had to rely on a heuristic model. The development of the carbonic anhydrase inhibitor dorzolamide is a nice example for the successful use of an experimentally determined 3D protein structure in the inhibitor design of an inhibitor which eventually led to a marketed drug.

The ocular disease glaucoma is caused by an excessive secretion of aqueous humor, the fluid that fills the anterior and posterior chambers of the eye. It has been known for some time that inhibition of the enzyme carbonic anhydrase reduces fluid secretion in the eye by blocking the conversion of carbon dioxide to bicarbonate.

The story of carbonic anhydrase began in 1932 with the discovery of the enzyme. The first inhibitors were found a few years later. For example, it was found that the sulfonamides discovered by Domagk also inhibit carbonic anhydrase. Phenylsulfonamide **15** (Fig. 15) binds to carbonic anhydrase with $IC_{50} = 300$ nM. In 1945, it was observed that heterocyclic molecules such as thiophene-2-sulfonamide **16** are even more potent inhibitors then the phenylsulfon-

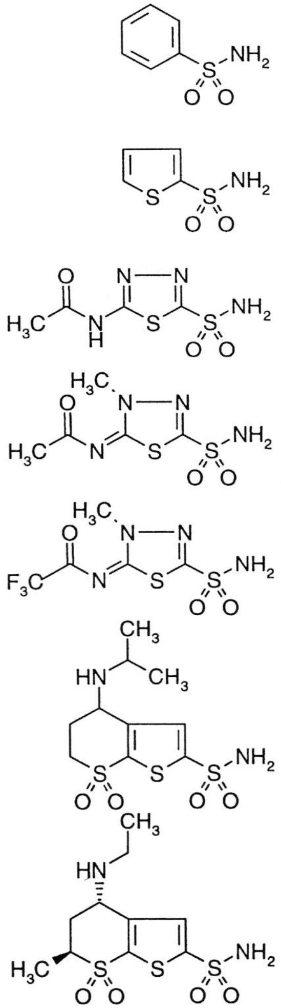

Figure 15. Chemical structures of the carbonic anhydrase inhibitors **15–21**. The small aromatic sulfonamides **15** and **16** bind with nanomolar affinity to carbonic anhydrase. Methazolamide **18** was used for a long time to treat glaucoma. **19** was the first topically active inhibitor. Structure-based drug design at Merck first led to **20** and then to the marketed drug, dorzolamide **21**.

amides. Following this discovery, a large number of heterocyclic sulfonamides were investigated. This work led to the discovery of acetazolamide (**17**) and methazolamide (**18**) as new carbonic anhydrase inhibitors. Methazolamide has been used for 40 years in the treatment of glaucoma. Unfortunately, the drug is not topically active, as it cannot penetrate ocular tissue. Systemic application of this drug is accompanied by side effects caused by the inhibition of carbonic anhydrase outside the eye. In 1983, the first topically active inhibitors were described. A strikingly small difference – a methyl group is replaced by a trifluoromethyl group – results in the topically active compound **19**. After this breakthrough, a large number of carbonic anhydrase inhibitors were investigated to determine the range of lipophilicity which will allow the compound to be topically active.

At Merck, the structure-based design of carbonic anhydrase inhibitors was started in the mid-1980s. The first compound with molecular modeling and X-ray structure determination playing an important role in the discovery process was thienothiopyrane-sulfonamide **20** (MK-927). This binds to carbonic anhydrase with a subnanomolar binding constant ($K_i = 0.7$ nM).

The 3D structure of carbonic anhydrase complexed with MK-927 was determined. As expected, the nitrogen atom of the sulfonamide group binds to the zinc ion in the active site of the enzyme. Furthermore, the inhibitor forms several hydrogen bonds and extensive lipophilic contacts with the protein. Somewhat surprisingly, however, it was found that the amino-isopropyl side chain adopts the energetically unfavorable axial position. Therefore, a modification which lowers the energy difference between the axial and the equatorial position of the side chain should increase the binding affinity. This effect can be obtained by introducing an additional methyl substituent at the six-membered ring. From the known 3D-structure it was obvious where an additional substituent could be appended without steric problems. The increased lipophilicity by the additional methyl group was compensated by shortening the amino-isopropyl side chain by on methyl group. The result of this modeling study was dorzolamide (**21**). This compound binds with $K_i = 0.37$ nM to carbonic anhydrase. The 3D structure of the carbonic anhydrase–dorzolamide complex is shown in Fig. 16. Dorzolamide has success-

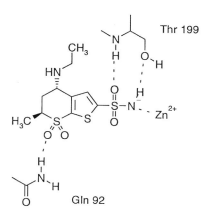
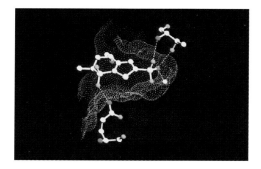

Figure 16. 3D structure of the complex of carbonic anhydrase with the inhibitor dorzolamide **21**. The sulfonamide group binds to the zinc cation and forms two additional hydrogen bonds with the enzyme. The sulfone group forms another hydrogen bond.

fully passed the clinical evaluation and is marketed since 1995 under the brand name Trusopt as the first topically active inhibitor of carbonic anhydrase for the treatment of glaucoma. In the first half of 1996, Trusopt recorded sales of 72 million US dollars. It has already become the most widely prescribed antiglaucoma product in the United States within its first year on the market. It is propably the first example of a marketed drug which originates from structure-based drug design using the experimentally determined 3D structure of the target enzyme.

2.4 Inhibitors of Serine Esterases

Serine hydrolases are enzymes that play a key role in diverse physiological systems. They all use a serine side-chain hydroxy group as a nucleophile in their enzymatic reaction. In contrast to the serine proteases of the trypsin/elastase family discussed above, the two esterases discussed here belong to a different mechanistic class that shares no sequence homology or structural similarity. Lipases digest nutritional fat triglycerides and acetylcholinesterase degrades the synaptic neurotransmitter, acetylcholine.

The following study has been undertaken in order to elucidate the distinctly different mode of action of the two esterases compared to trypsin, to rationalize structure–activity relationships of known inhibitors, and to form a basis for future ligand design projects.

2.4.1 Human Pancreatic Lipase (hPL)

Human pancreatic lipase is a protein consisting of 445 amino acids. It folds in two domains with the larger N-terminal containing the catalytic site. The overall folding is represented in a ribbon-diagram in Fig. 17. This crystallographic structure of the human pancreatic lipase represents an inactive form of the enzyme [17]. The active site is totally buried by a loop and

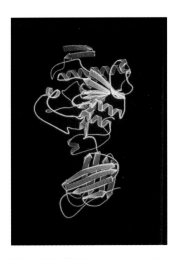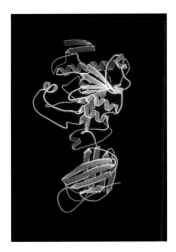

Figure 17. Ribbons representation of the hPL structure. The upper catalytic domain consists of a central β-sheet with helices on both sides. The loop covering the active site is the two turn helix at the left side. The bottom domain consists of two β-pleated sheets.

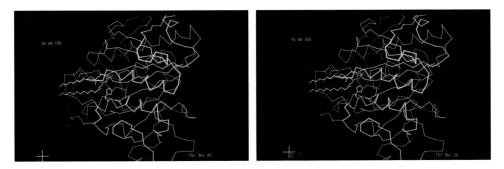

Figure 18. Cα-display of the hPL structure with the catalytic triad in red, the loop covering the active site in pink, beginning and ending at a disulfide bridge in yellow. A triglyceride substrate molecule in green is attached to the serine-Oγ. Clearly, the loop interferes with substrate binding and has to be removed by a movement to the left before the reaction takes place.

the catalytic serine is inaccessible to solvent and substrate (Fig. 18). The enzyme has to undergo interfacial activation [18, 19] at the surface of a fat droplet and thereby exposes the hydrophobic interior of this loop to the fat; the active site becomes accessible.

2.4.2 Model of the Trilaurin Triglyceride Substrate Binding

Since the structure observed in the crystals represents an inactive form of the enzyme, several structural modifications were required to accommodate a substrate molecule in the active site. These modifications were performed such that the resulting model remained as similar as possible to the experimental structure. First, the loop covering the active site which is anchored at a disulfide bridge was shifted to an alternative strainless conformation opening up the active site. Second, the side-chain orientations of three aromatic residues near the active site had to be changed (Fig. 19). Third, the segment to the right of the active site had to be

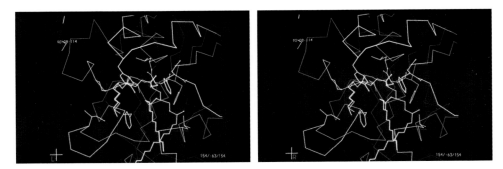

Figure 19. Structure of the 'active model' of hPL with some selected side chains in yellow. The top three aromatic side chains have altered conformations, the segment containing Phe77 at the right has been shifted by 1.5 Å away from the catalytic serine. The model of the substrate transition state is displayed in pink, the resulting product ester in red. The conformation of the fatty acid chains has been arbitrarily chosen to be extended.

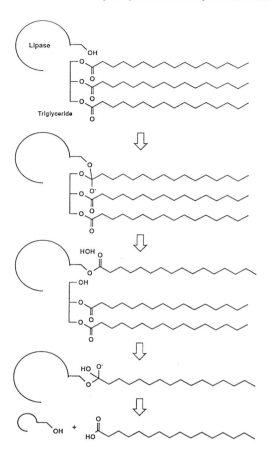

Figure 20. Reaction scheme of the triglyce-ride cleavage by pancreatic lipase.

shifted away from the catalytic serine by about 1.5 Å. The model created by these three steps was then used to dock the substrate trilaurin triglyceride starting from the X-ray structure of the molecule alone [20] with the following requirements in mind: the trilaurin ester carbonyl comes in contact with the side-chain oxygen of the catalytic serine; the ester carbonyl oxygen sits in the oxy-anion hole formed by backbone NH groups; the ester is in the *trans*-conforma-tion and the fatty acid tails protrude to where the fat droplet is assumed to be located. Start-ing from this model of the Michaelis complex, the consecutive catalytic steps (Fig. 20) were built (the hemiacetal model is shown in Fig. 19). These models represent a reaction trajectory with essentially strainless intermediates. The backprotonation of the leaving alcohol group and the activation of the incoming water could be performed by the neighboring histidine Nε.

2.4.3 Tetrahydrolipstatin (THL)

THL (**22**), a hydrogenated derivative of lipstatin originally isolated from *Streptomyces toxy-tricini*, is a potent inhibitor of hPL [21, 22] and is currently being evaluated in clinical trials.

The model (Figs. 21 and 22) derived from the previously discussed substrate model suggests that the tetrahedral hemiacetal intermediate of the β-lactone decays though ring opening. The hydroxy ester formed is conformationally restricted by the protein environment in the narrow cleft below the catalytic serine. The hydroxy group thus protects the ester group from incoming deacylating water. The enzyme remains acylated and therefore inactivated. The hydrophobic moieties of THL, the two hydrocarbon chains on the lactone ring and the leucine side-chain coincide with the three fatty acid chains in the trilaurin model which are thought to protrude into the fat droplet.

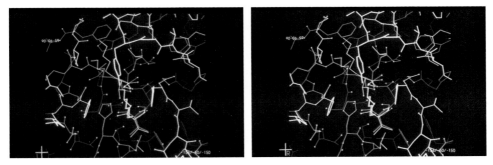

Figure 21. Reaction scheme of the inhibition of pancreatic lipase by THL (**22**). The reaction stops at the acyl–enzyme intermediate because the alcohol group prevents the attack by a deacylating water molecule.

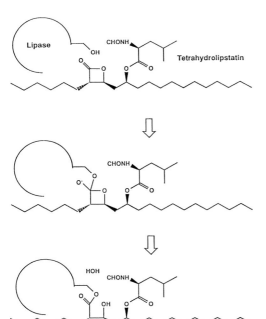

Figure 22. Model of the THL (**22**) transition state in the active hPL in red and the model of the product in yellow. The alcohol group is locked in a position which prevents a water molecule from attacking the ester for deacylation.

This 3D model rationalizes the experimentally observed formation of a hydrolytically stable acyl enzyme when β-lactones of the THL class bind to lipases.

2.5 Acetylcholinesterase (AChE)

The X-ray structure of AChE [23] of the electric eel *Torpedo californica* reveals a single domain fold with a characteristic central eleven strand β-sheet. Although no detectable homology is present between hPL and AChE, the folding in the core of the catalytic domain of the lipase is very similar to that of AChE (Fig. 23). The central β-sheet can be superimposed such that the long helix which carries the catalytic serine at its N-terminus aligns. The geometry of the active site functional groups is also similar, but there are distinct differences. The active site of AChE is located at the bottom of a deep gorge open to solvent. This gorge is coated with aromatic side chains. The catalytic triad Ser-His-Glu is similar in geometry, but the His and Glu side chains are attached to different secondary structure elements than in hPL [24]. The oxy-anion hole consists of two equivalent NH-groups (Fig. 24). In contrast to the hPL

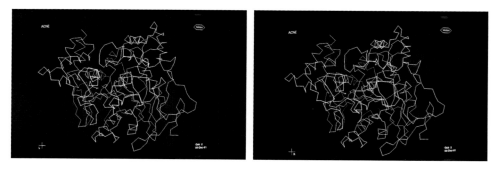

Figure 23. Cα-display of the crystal structure of AChE in yellow with the central sheet in blue, the catalytic triad in green, and the central helix starting at the catalytic serine in pink.

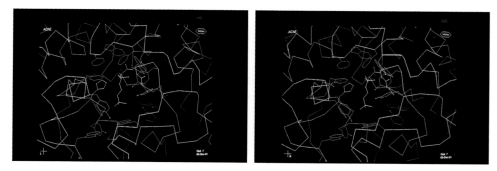

Figure 24. Close-up view of the active site of AChE with the triad in yellow, the aromatic side chains coating the deep gorge in red, and a model of the acetylcholine transition state in pink.

structure, the AChE structure appears to represent the active form of the enzyme with appropriate geometry of the active site residues. Gratifyingly, the geometry of the active site in the hypothetical model of the active form of the lipase (which had been built prior to the availability of the AChE structure) is very similar to AChE.

Several structures of small molecule complexes with acetylcholinesterase have been solved. They reveal a binding site next to the catalytic serine preferrentially occupied by a positively charged moiety next to a hydrophobic portion. The positively charged functional groups almost superimpose in front of a tryptophan residue at the bottom of the gorge [25–27].

2.5.1 Model of the Acetylcholine Substrate Binding

Starting from the hypothetical model of the active hPL superimposed to the AChE structure, a model of the AChE reaction mechanism of acetylcholine **23** hydrolysis could be built (Figs. 25 and 26). In the hemiacetal transition state, the OH of the choline leaving group is at hydrogen bonding distance from the Nε of the catalytic histidine. The choline molecule is accommodated in the fully extended conformation and positions the trimethylammonium group towards the face of the indole system of Trp84. The methyl group of the acetyl moiety is placed in a small pocket which could not accommodate larger or branched acid components. This is in agreement with the observed reduced hydrolysis rates of choline homologs [28]. The acetylated serine can be attacked by a water molecule which follows the reverse trajectory of the leaving alcohol group. The water molecule can be activated by transfer of a proton to the Nε of the catalytic histidine. A superposition of the transition state models of hPL with trilaurin and AChE and ACh is shown in Fig. 26. The functional groups of the enzyme involved in the catalysis occupy similar positions. The acid moiety of the triglyceride extends through a narrow channel in the lipase while there is no such channel in AChE. The leaving diglyceride in the lipase extends to the protein surface while in the AChE the choline fills the bottom of the gorge.

2.5.2 Physostigmine

The alkaloid physostigmine **24** from calabar bean [29] inactivates AChE through transfer of a carbamoyl group to the catalytic serine; this serine carbamate is only slowly hydrolyzed. The molecule (Fig. 27) has very little flexibility. Starting from the X-ray structure of the molecule alone [30], physostigmine was docked to the AChE active site such that the catalytic serine could be carbamoylated. The only conformation fitting the active site has the basic nitrogen near the trimethylammonium group in the substrate model. The tricyclic system essentially fills the bottom of the gorge (Fig. 28). The N-methyl group of the carbamate protrudes into the restricted pocket which accommodates the acetyl group in the substrate model; larger substituents at this position would require a different, more strained conformation. This is in agreement with the structure–activity relationship found for such a series of compounds [31].

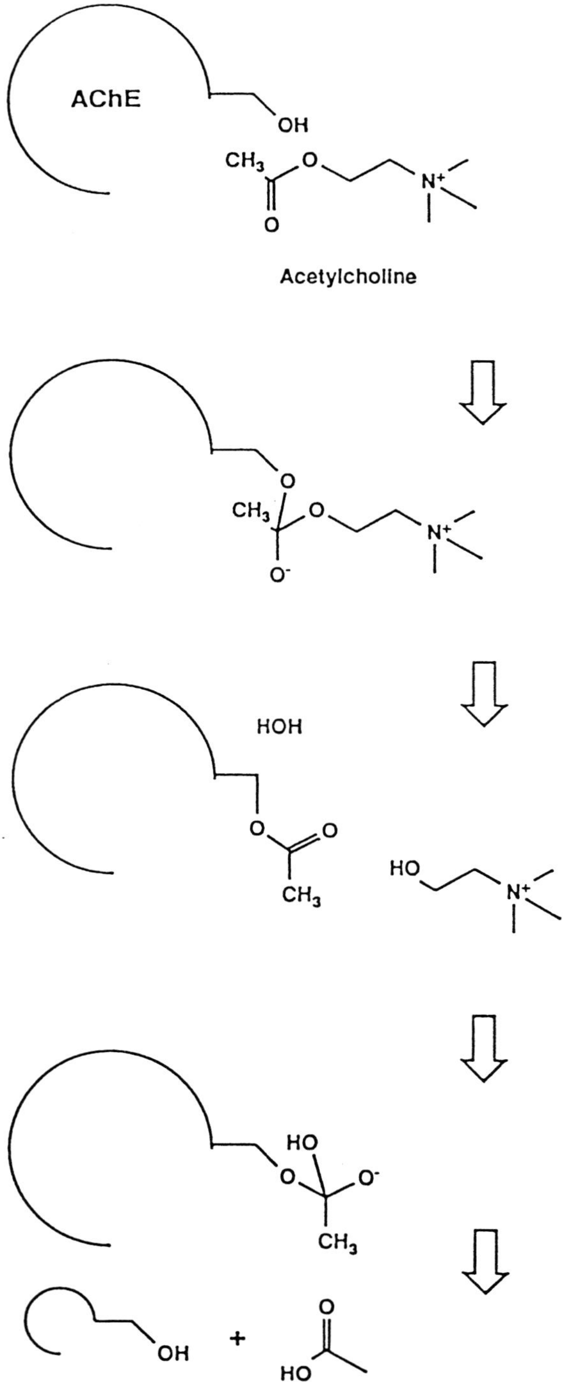

Acetylcholine

Figure 25. Reaction scheme of the cleavage of acetylcholine **23** by acetylcholinesterase.

Figure 26. Superposition of active site and substrate models of AChE in yellow with a pink substrate and of lipase in blue with a green substrate. The superpositioning was done with respect to five C-α positions in each of the five central β sheets.

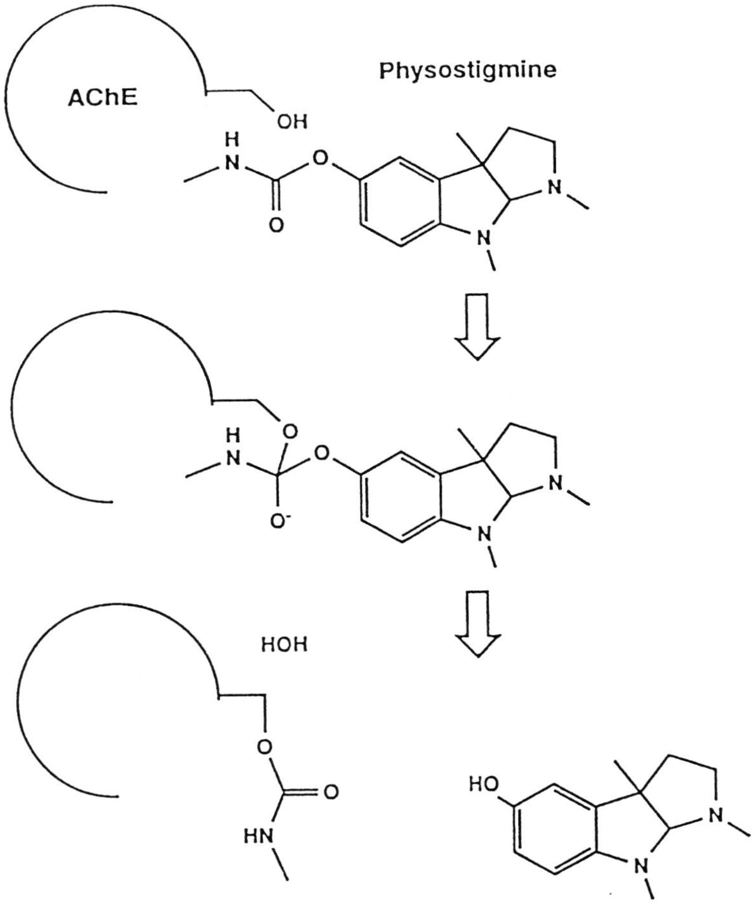

Figure 27. Reaction scheme of the carbamoylation reaction of physostigmine **24** with acetylcholinesterase.

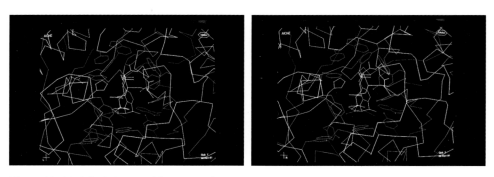

Figure 28. Model of the transition state with physostigmine (**24**, green) in AChE; the acetylcholine model is shown in pink for comparison.

2.5.3 Eisai E2020

The AChE inhibitor E2020 (**25**) developed at Eisai is a representative of a large series of very potent compounds [32–34]. Initial attempts to position the indanone carbonyl group of E2020 close to the catalytic serine were unsuccessful; the compound had to be folded and could then not fit the gorge. Positioning the basic nitrogen close to the corresponding position in the other models allows only one orientation of the molecule: The indanone portion fills the gorge towards the solvent while the benzyl group fills a cavity at the bottom of the gorge. The model shown in Fig. 29 has exactly the same conformation with respect to Θ_1 and Θ_2 (Fig. 30) that had been deduced from an extensive QSAR study [33]. The conformation of the benzyl group is different and adopts an endo orientation in order to reach the cavity. The QSAR data seem to confirm this model: a wide variety of substituents is tolerated at different positions of the indanone aromatic ring; in the model this portion is directed toward the exterior. In contrast, substitution at the para postion of the benzyl group reduces activity dramatically; also benzoyl instead of benzyl is not tolerated. This can be rationalized from the model showing that the benzyl group essentially fills up the cavity in the conformation shown.

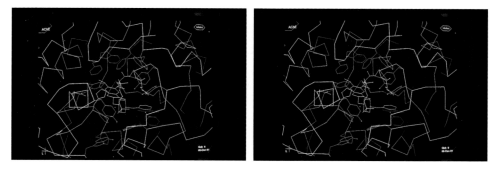

Figure 29. Model of the complex of E2020 (**25**, green) with AChE, again with acetylcholine in pink.

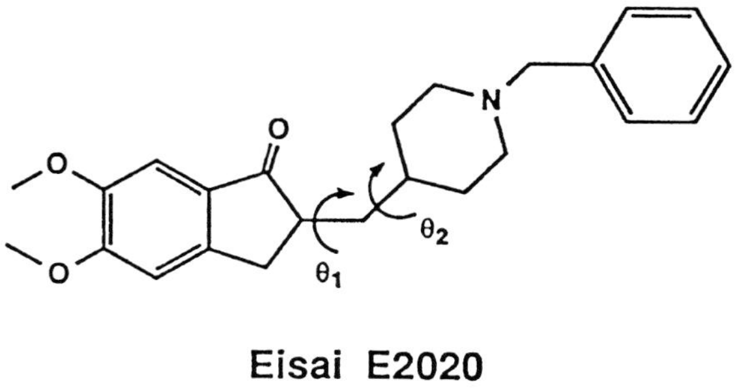

Eisai E2020

Figure 30. Formula of the E2020 AChE inhibitor **25** with two torsional angles indicated.

These results suggest that E2020 is a noncovalent AChE inhibitor which binds to the hydrophobic gorge but does not interact with the catalytic residues.

This study demonstrates that mechanistic insights can be applied to families of enzymes once convincing models have been derived for the first member. The combined use of structure–activity relationship and structural arguments form the basis projects aiming at improved inhibitors.

References

[1] Huber, R., Kukla, D., Bode, W., Schwager, P., Bartels, K., Deisenhofer, J., and Steigemann, W., *J. Mol. Biol.* **89**, 73–101 (1974)

[2] Banner, D. W., and Hadvary, P., *J. Biol. Chem.* **266**, 20085–20093 (1991)

[3] Stubbs, M. T., Oschkinat, H., Mayr, I., Huber, R., Angliker, H., Stone, S. R., and Bode, W., *Eur. J. Biochem.* **206**, 187–195 (1992)

[4] Feng, N., Konishi, Y., Frazier, R. B., and Scheraga, H. A., *Biochemistry* **28**, 3082–3094 (1989)

[5] Martin, P. D., Robertson, W., Turk, D., Huber, R., Bode, W., and Edwards, B. F. P., *J. Biol. Chem.* **267**, 7911–7920 (1992)

[6] Banner, D. W., Ackermann, J., Gast, A., Gubernator, K., Hadvary, P., Hilpert, K., Labler, L., Mueller, K., Schmidt, G., Tschopp, T., van de Waterbeemd, H., and Wirz, B. Serine Proteases: 3D Structures, Mechanism of Action and Inhibitors. In: Testa, B., Kyburz, E., Fuhrer, W., Giger, R. (Eds.), *Perspectives in Medical Chemistry*, 27–43, VHCA, Basel (1993)

[7] Banner, D. W., Ackermann, J., Gast, A., Gubernator, K., Hadvary, P., Hilpert, K., Labler, L., Mueller, K., Schmid, G., Tschopp, T., van de Waterbeemd, H., and Wirz, B., *J. Med. Chem.* **37**, 3889–3901 (1994)

[8] Schmid, G., Ackermann, J., Banner, D., Gast, A., Gubernator, K., Hadvary, P., Hilpert, K., Labler, L., and Tschopp, T., *J. Pharm. Belg.* **50**, 188–193 (1995)

[9] Schmid, G., Ackermann, J., Banner, D. W., Gubernator, K., Hadvary, P., Hilbert, K., Labler, L., Mueller, K., Tschopp, T., Wessel, H. P., and Wirz, B., *European Patent 0 468 231* (1990)

[10] Ackermann, J., Banner, D. W., Gubernator, K., Hilpert, K., and Schmid, G., *European Patent 559 046* (1992)

[11] Veale, C. A., Bernstein, P. R., Bryant, C. et al., *J. Med. Chem.* **38**, 98–108 (1995)

[12] Dolle, R. E., Prouty, C. P., Prasad, C. V. C., Cook, E., Saha, A., Ross, T. M., Salvino, J. M., Helaszek, C. T., and Ator, M. A., *J. Med. Chem.* **39**, 2438–2440 (1996)

[13] Semple, J. E., Rowley, D. C., Brunck, T. K., Ha-Uong, T., Minami, N. K., Owens, T. D., Tamura, S. A., Goldman, E. S., Siev, D. V., Ardecky, R. J., Carpenter, S. H., Ge, Y., Richard, B. M., Nolan, T. G., Hakanson, K., Tulinsky, A., Nutt, R. F., and Ripka, W. C., *J. Med. Chem.* **39**, 4531–4536 (1996)

[14] Greer, J., Erickson, J. W., Baldwin, J. J., and Varney, M. D., *J. Med. Chem.* **37**, 1035–1052 (1994)

[15] Baldwin, J. J., Ponticello, G. S., Anderson, P. S., *J. Med. Chem.* **32**, 2510–2513 (1989)

[16] Cushman, D. W., Cheung, H. S., Sabo, E. F., and Ondetti, M. A., *Biochemistry* **16**, 5484–5491 (1977)

[17] Winkler, F. K., D'Arcy, A., and Hunziker, W., *Nature* **343**, 771–774 (1990)

[18] Gubernator, K., Winkler, F. K., and Mueller, K., The structure of human pancreatic lipase suggests a locally inverted, trypsin-like mechanism. Proceedings of the CEC-GBF Workshop 1990 in Braunschweig: *Lipases: Structure, Mechanism and genetic Engineering*, Alberghina, L., Schmid, R. D., Verger, R. (Eds.), GBF Monographs, VCH Publishers, Weinheim, pp. 9–16

[19] Winkler, F. K., and Gubernator, K., Structure and mechanism of human pancreatic lipase. In: *Lipases*, Woolley, P., Petersen, S. B. (Eds.). Cambridge Univ. Press, Cambridge, UK, 139–157 (1994)

[20] Gibon, V., Blanpain, P., Norberg, B., and Durant, F., *Bull. Soc. Chim. Belg.* **93**, 27 (1984)

[21] Weibel, E. K., Hadvary, P., Hochuli, E., Kupfer, E., and Lengsfeld, H., *J. Antibiot.* **40**, 1081–1085 (1987)

[22] Luethi-Peng, Q., Maerki, H. P., and Hadvary, P., *FEBS Lett.* **299**, 111–115 (1992)

[23] Sussman, J. L., Harel, M., Frolow, F., Oefner, Ch., Goldman, A., Toker, L., and Silman, I., *Science* **253**, 872–879 (1991)

[24] Schrag, K. D., Winkler, F. K., and Cygler, M., *J. Biol. Chem.* **267**, 4300–4303 (1992)

[25] Harel, M., Schalk, I., Ehret-Sabattier, L., Bouet, F., Goeldner, M., Hirth, C., Axelsen, P., Silman, I., and Sussman, J., *Proc. Natl Acad. Sci. USA* **90**, 9031–9039 (1993)

[26] Villalobos, A., Blake, J. F., Biggers, C. K., Butler, T. B., Chapin, D. S., Chem, Y. L., Ives, J. L., Jones, S. B., Liston, D. R., Nagel, A. A., Nason, D. M., Nielsen, J. A., Shalaby, I. A., and White, W. F., *J. Med. Chem.* **37**, 2721–2734 (1994)

[27] Wlodek, S. T., Antosoewicz, J., McCammon, J. A., Straatsma, T. P., Gilson, M. K., Broggs, J. M., Humblet, C., and Sussman, J. L., *Biopolymers* **38**, 109–117 (1996)

[28] Quinn, D. M., *Chem. Rev.* **87**, 955–979 (1987)

[29] Engelhart, E. O., *Loewi Nauyn-Schmiedeberg's* **150**, 1 (1930)

[30] Pauling, P., and Petcher, T. J., *J. Chem. Soc. Perkin* **2**, 1342 (1973)

[31] Harris, L. W., Anderson, D. R., Pastelak, A. M., Bowersox, S. L., Vanderpool, B. A., and Lennox, W. J., *Drug Chem. Toxicol.* **15**, 127–143 (1992)

[32] Cardozo, M. G., Iimura, Y., Sugimoto, H., Yamanishi, Y., and Hopfinger, A. J., *J. Med. Chem.* **35**, 584–589 (1992)

[33] Cardozo, M. G., Kawai, T., Iimura, Y., Sugimoto, H., Yamanishi, Y., and Hopfinger, A. J., *J. Med. Chem.* **35**, 590–601 (1992)

[34] Yamanishi, Y., Ogura, H., Kosasa, T., Arika, S., Sawa, Y., Yamatsu, K., *Adv. Behav. Biol.* **38B**, 409–413 (1990)

3 Structure-Based Design: From Renin to HIV-1 Protease

Elizabeth A. Lunney and Christine Humblet

Abbreviations

Abbreviations follow the recommendations of the IUPAC-IUB Joint Commission on Biochemical Nomenclature for amino acids and peptides: *Eur. J. Biochem.* **158**, 9–31 (1984).

3D	3-Dimensional
NMR	Nuclear Magnetic Resonance
Smo	4-Morpholinesulfonic acid
Alg	S-Allylglycine
HIV-1	Human Immunodeficiency Virus (Type 1)
AIDS	Acquired Immuodeficiency Syndrome
RSV	Rous Sarcoma Virus
Boc	*tert*-butoxycarbonyl
Atm	3-(2-amino-4-thiazolyl)alanine
ACHPA	(3*S*,4*S*)-4-amino-3-hydroxy-5-cyclohexylpentanoic acid
ACDMH	(2*S*,3*R*,4*S*)-2-amino-1-cyclohexyl-3,4-dihydroxy-6-methylheptane
Ets	S-Ethylthioglycine
AMPMA	*m*-bis(aminomethyl)benzene

3.1 Introduction

Approaches to design enzyme inhibitors have commonly relied on the amino acid composition of the known peptide substrates. An iterative process ensued to convert the regular peptide substrates or inhibitors using concepts of enzyme mechanisms and peptide, or peptidomimetic chemistry. Over the past decade, the inhibitor design process has been impacted increasingly by the availability of 3D structural representations of the targeted enzyme binding region. Ideally, the enzyme structures originate from experimental data such as X-ray crystallography or NMR spectroscopy. Offering a less accurate alternative, homology-based models can also be assembled from structurally related templates. Thus, the challenge of designing substrate-based nonpeptide inhibitors now benefits from technological advances, which recently occurred in computational methods, spectroscopy and biochemistry. These have allowed for structure-based design to become an important and powerful companion in inhibitor design [1–6].

Structure-based design approaches have been extensively explored with the aspartic protease family of enzymes. A target in antihypertension treatment, renin was among the early aspartic protease candidates for structure-based design. As will be described, this ambitious, early undertaking provided critical ground work for later applications. Thus, today, we are witnessing the impact that the renin structure-based design has had in ongoing work with the HIV-1 protease in the AIDS research.

3.2 Renin

Renin is an enzyme that processes angiotensinogen to produce the decapeptide angiotensin I, a precursor of the vasoconstrictor, angiotensin II [7]. Inhibiting renin's function could interrupt the renin–angiotensin cascade and thus, possibly provide a successful way to treat hypertension. An advantage in targeting renin lies in the high selectivity that the enzyme demonstrates for its natural substrate, angiotensinogen.

At the time that renin inhibitor research began in our laboratories, no 3D renin structure was available to aid in the endeavor. However, X-ray crystal structures of various homologous fungal aspartic proteases were being determined [8–10]. These data, combined with the known primary sequences of aspartic proteases, were used to construct models of mouse and human renin [11–13]. The models were more particularly refined to address the molecular details of the binding site in the region of the catalytic aspartic acids. In our laboratories, the X-ray crystal structure of the fungal enzyme, endothiapepsin, derived from *Endothia parasitica*, served as a template for the renin model used in inhibitor design (Fig. 1) [14].

The crystal structures of aspartic proteases helped elucidate the mechanism of action for this family of enzymes. For example, in the crystal structure of native endothiapepsin, a water

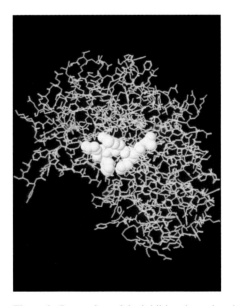 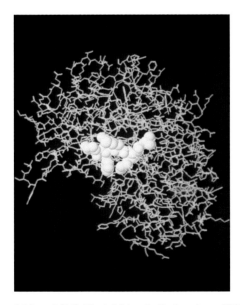

Figure 1. Stereoview of the inhibitor-bound renin model (cyan) [14]. The inhibitor is displayed as a CPK model (yellow).

molecule is oriented between the two nearly co-planar catalytic aspartic acid carboxylates [15]. This catalytic water is postulated to participate in the nucleophilic attack at the substrate amide bond [16]. The crystal structures of the aspartic proteases bound with various inhibitors unravelled molecular details of the binding of the peptide-like inhibitors, the majority of which were transition state analogs of the substrate hydrolysis process (Fig. 2). Thus, the scissile amide isosteres displace the catalytic water observed in the native structure as they bind to the aspartic dyad.

The surrounding binding region includes a flexible hairpin loop fragment, commonly referred to as the flap. As additional crystal structures were determined, it became apparent that the flap movement was likely to participate in inhibitor binding [10]. As the inhibitors extend to the amino and carboxy termini, the amide functionalities form hydrogen bonds with the flap region, and on the opposite side, with various residues contained mainly in the C-terminal domain of the enzyme. The hydrogen bonding appears to provide a zippering framework on either side of the aspartic dyad (Scheme 1) [17]. Upon inhibitor binding, a dynamic movement or a rigid body shift in the enzyme binding was noted in the crystallography data obtained for endothiapepsin–inhibitor complexes [18]. However, it was observed that an inhibitor with a truncated amino-terminus did not show the rigid body movement in the enzyme in a nonisomorphous cell unit, which occurred with other inhibitors that have these residues. Therefore, the domain movement may only occur with an inhibitor having a certain minimum length and may be essential for the potent binding.

The residue side chains occupy subsites on alternate sides of the backbone, designated according to the nomenclature introduced by Schechter and Berger (P1, P1′ ... , S1, S1′) [19]. These experimental observations guided the design of inhibitor structures tailored to mimic

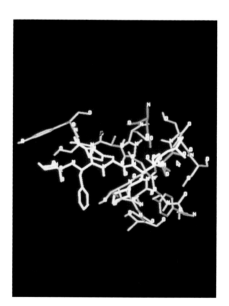 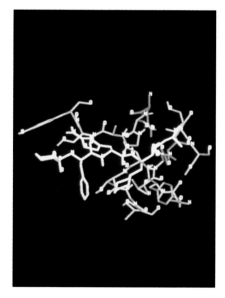

Figure 2. Stereoview of a renin inhibitor, Smo-Phe-Atm-ACDMH (yellow), bound in the renin model binding site (green, catalytic aspartic acids in red).

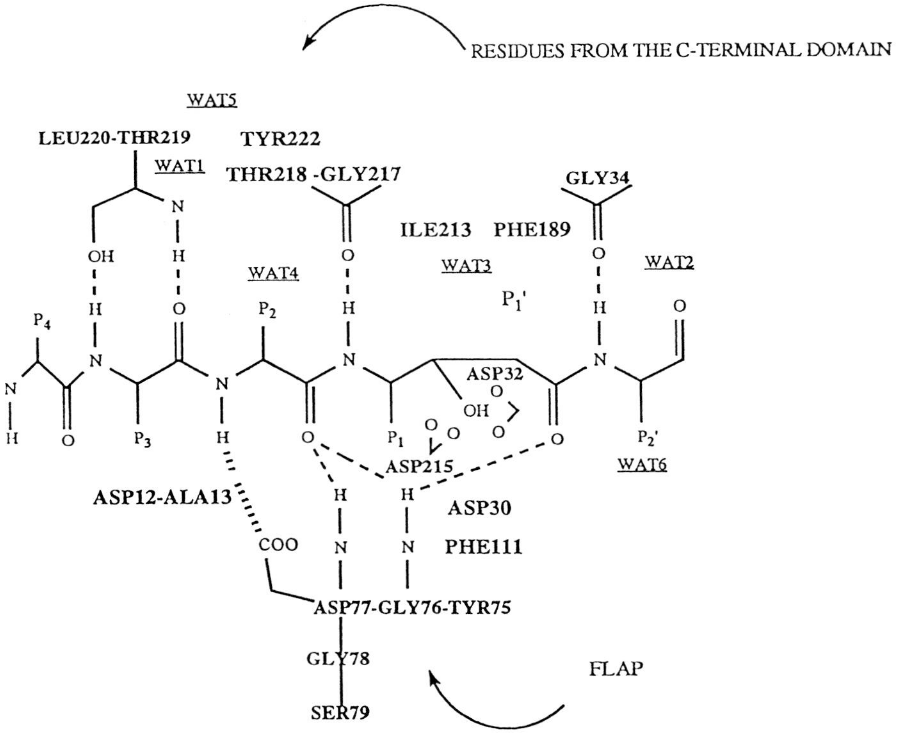

Scheme 1. Hydrogen bonding scheme observed in five X-ray structures of endothiapepsin bound with renin inhibitors [17]. (||||) Non-conserved hydrogen bond; (---) conserved hydrogen bonds.

the known extended peptide-like bound conformations. Target inhibitors were docked in the renin active site and evaluated for steric contacts, polar interactions and overall compatibility with the enzyme binding cavity. The rational design approaches applied concepts from peptide mimetics, transition-state replacements, and structural elements extracted from known crystal structures. In the absence of well-established modeling or computational protocols, the docking experiments remained manual, often user-biased, and mostly qualitative. Concurrent with optimizing the binding interactions, a basic strategy behind the inhibitor design was to diverge from the peptidic nature of the known potent inhibitors and at the same time reduce their molecular weight. These features have been associated with poor pharmacokinetic behavior, including metabolic instability, poor absorption, and/or high hepatic clearance [7, 20]. It was reasoned that the replacement of the peptidic elements could ultimately enhance the oral bioavailability of the proposed compounds. Toward this end, unnatural amino acids, amide bond replacements and constrained moieties were incorporated into structure-based inhibitor design. The inhibitor-bound complex was scrutinized to capitalize to the fullest on key, potent interactions, while simultaneously evaluating the removal of less productive functionalities in the ligand. These modifications were carried out in an incremental manner focusing on individual residues of the peptide-like inhibitor. As illustrated below, the ability to modify the peptidic nature and to reduce the molecular size of potent renin inhibitors, and at the same time retain activity, was significantly limited and led to many unexpected results. The latter often prompted retrospective reassessments of the models, and spurred the development and application of computational methodologies, which ultimately enhanced newer design strategies.

3.2.1 Catalytic Site Binding

Initial inhibitors were designed to incorporate transition state analogs of the substrate scissile moiety. One of the first sites analyzed with molecular modeling was the catalytic dyad, where the X-ray structures clearly indicated stringent binding requirements involving bifurcated hydrogen bonds. The 3D structure of the fungal aspartic protease, *Rhizopus chinensis*, bound with pepstatin provided critical insight that helped elucidate the binding mode of the naturally occurring transition state moiety, statine (Fig. 3) [21, 22]. From these studies, it was concluded that statine was a dipeptide mimetic, a fundamental discovery in early renin research. Statine was subsequently optimized incorporating, for example, a larger cyclohexyl substituent, which more completely filled the S1 site and resulted in improved potency.

Over the years, the 3D studies of P1-P1' isosteres for renin inhibitors evolved to the design of potent diols [23–25] (**1** and **2**, Table 1). The P1' hydroxyl in the P1-P1' diol had been predicted to bind to one of the aspartic acids in the catalytic dyad [25]. Subsequent structure–activity relationships observed for an azidoglycol series indicated that the P1' hydroxyl was actually a hydrogen acceptor [26]. Further studies involved detailed molecular modeling and molecular dynamics simulations carried out with an enzyme model refined on the basis of the newly released crystal structure of porcine pepsin. The results supported an interaction between the P1' hydroxyl group and Ser76(NH) in the flap region. This binding mode has since been further substantiated in the X-ray structure of a diol inhibitor bound in endothiapepsin, which has a Gly in place of the Ser at position 76 [17] (Fig. 4).

Analogs with a one-four diol isostere, for example **3**, were also found to have high affinity [17].

3.2.2 Backbone Variations

The models of the renin structure indicated that the P2 subsite was narrow at the backbone region, so disubstitution of the P2 α-carbon was not recommended. Subsequent experimental evidence confirmed this prediction. In one particular series, where epimerization of the aminomalonate at the P2 site of inhibitor **4** (Table 1) was a concern [27], the α-methyl deriv-

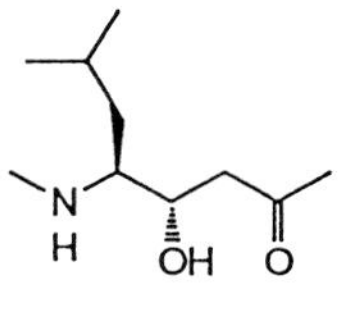

Figure 3. Statine – A transition state mimetic replacing a dipeptide.

Table 1. Renin inhibitors

	IC$_{50}$ (nM)

1.   5

2.   Boc[a]-NH ...   1.5

3.[b]   1.7

4.[b]   0.28   [a] Boc = *tert*-butoxycarbonyl
                                                                                    [b] Mixture of diastereomers

ative was synthesized and tested. The activity dropped dramatically (4.4% at 10^{-8} M) relative to the parent analog [17], thus strongly supporting the above model-based prediction.

Through both evaluating endothiapepsin crystal complexes and analyzing inhibitor interactions in the enzyme binding cleft of the renin model, the structure–activity relationship for a series of analogs containing P3-P2 amide isosteres and Gly at P2 (Table 2) was rationalized [28]. This work suggested that the hydroxyethylene isostere in **5** could function as a hydrogen bond acceptor with the enzyme in place of the standard amide carbonyl. The observation was

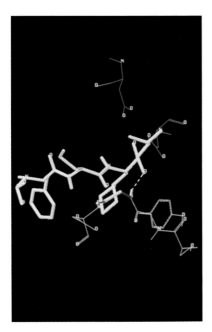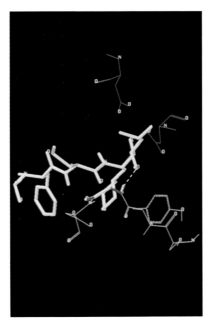

Figure 4. Stereoview of the X-ray crystal structure of a renin inhibitor, Smo-Phe-Ets-ACDMH (yellow), bound in endothiapepsin [17]. The hydrogen bond between the P1′ hydroxyl and Gly76(NH) (magenta) is shown by a white dotted line. The catalytic aspartic acids are shown in red.

subsequently supported by a crystal structure of the inhibitor bound to endothiapepsin (Fig. 5), which clearly showed the interaction between the hydroxyl oxygen and Thr219(NH) in the binding site [29]. The analog having the double bond isostere with the 'E' geometry, **6**, was compatible with the favored extended conformation of the potent inhibitors, whereas **7**, with the 'Z' configuration was not. This correlated with the activity data that showed micromolar binding with the former, and no activity with the latter. The series of inhibitors also proved that a P2 side chain was not essential for potent binding, as is illustrated by **5**.

Structure-based design is an evolving process, which seeks to incorporate new computational and experimental data as they become available. Thus, incorrect binding predictions often led to additional studies to understand the pitfalls of the originally proposed hypotheses. This was clearly illustrated in an analog design involving constrained P4-P3 moieties, such as in **8** [30]. In this series, a substituted piperidone ring system was incorporated at the P4-P3 site (Table 2). The 4S diastereomer was predicted to be the more active analog. Not only did the prediction fail but both diastereomers were less active than the acyclic analogs. In an effort to reinvestigate the modeling results, the Cambridge Crystallographic Database [31] was screened to retrieve known conformations for the piperidone ring and related cyclohexenes. Analog **8** was remodeled accordingly. Upon docking and minimizing these models in the renin binding site, it was observed that either a distortion of the 3S center in the ring occurred or the phenyl group was projected toward solvent. Either situation would be unfavorable to binding and thus, would help explain the loss in activity of the cyclic inhibitors relative to the

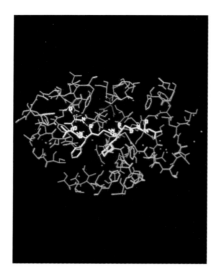 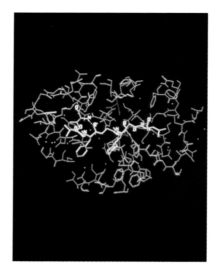

Figure 5. Stereoview of the X-ray crystal structure of a renin inhibitor, Boc-Phe(CHOHCH₂)Gly-ACHPA-LEU-AMPMA (yellow), bound in endothiapepsin (cyan, catalytic aspartic acids in red) [17]. A hydrogen bond between the hydroxyl oxygen and the Thr219(NH) (magenta) is shown by a white dotted line.

acyclic control compounds. The additional modeling revealed the conflict between lactamization through the P3 side chain and maintaining the extended conformation for the P4-P3 residues. This work clearly points out how incomplete conformational sampling can result in erroneous conclusions regarding ligand/protein interactions.

3.2.3 Subsite Interdependencies

As was noted earlier, the inhibitor-bound peptides adopt an extended backbone conformation. The amino acid side chains alternate binding on either side of the backbone, bringing every other side chains in close proximity (Fig. 2). This observation led to novel ligand design. For example, the close orientation of the P1′ and P2 side chains led to the design of cyclic compounds bridging these two subsites (Table 3, **9**) [32]. While occupying two adjacent pockets, it was hoped that the cyclic nature of the analog might improve the binding affinity due to a favorable entropic effect, and also increase metabolic stability relative to acyclic derivatives. The resulting inhibitor exhibited only weak binding. Other inhibitors arising from a similar strategy produced constrained analogs with nanomolar activities, as represented by **10** (Table 3) [33, 34]. A 3-quinuclidinone derivative of **10** inhibited plasma renin activity *in vivo* and also exhibited bioavailability. The modeling approaches applied in these design studies involved different depths of analysis. In the former case, the bridging moiety was tailored on the known linear peptide-bound backbone conformation; the resulting analog, **9**, was manually docked in the renin model. Limited energy minimization cycles were applied to assess or prevent major intermolecular incompatibilities. The poor binding observed subse-

Table 2. Renin inhibitors modified at the P3–P2 amide bond

IC$_{50}$ (nM)

5.[a] 22.0

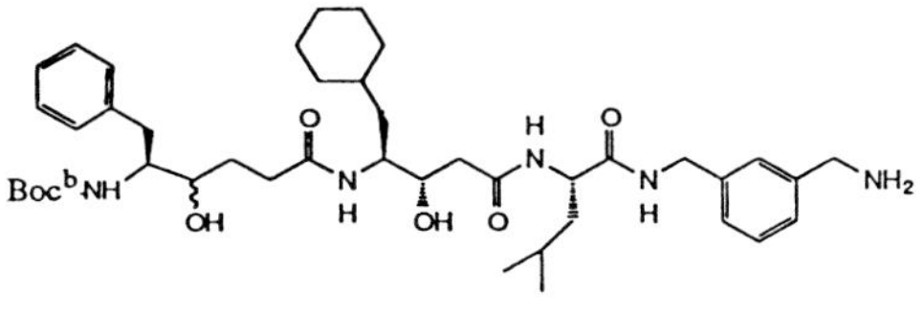

6. 7,000

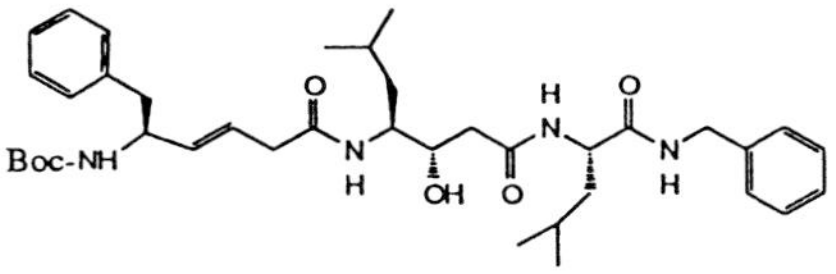

7. >100,000

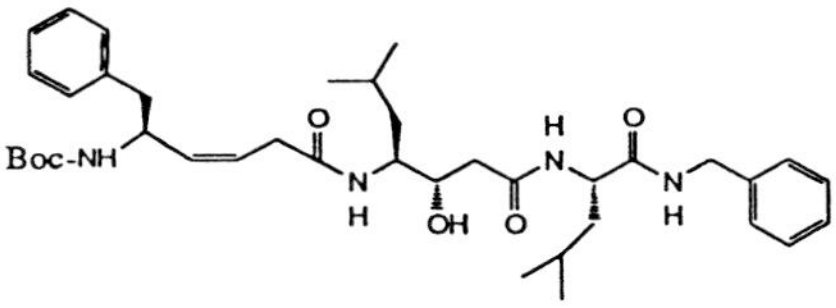

8. (4R) 134
(4S) 5910

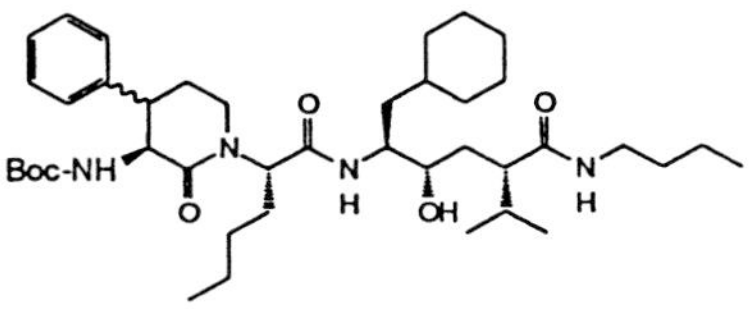

[a] Mixture of diastereomers
[b] Boc = *tert*-butoxycarbonyl

Table 3. Cyclic renin inhibitors

	IC_{50} (nM) or % at μM
9. BNMA[a]-HN...	26%
10. Boc[b]-NH...	3.4
11.	0.28
12.	6.7
13. Boc-NH...	20

[a] BNMA = bis-[(1-naphthyl)methyl]acetic acid
[b] Boc = *tert*-butoxycarbonyl

quently prompted further studies. These included a solution NMR assessment of the inhibitor conformational preferences. From this learning experience, the modeling protocol was revised to include conformational studies and internal energetics applied to the inhibitor structure both in the absence and the presence of the enzyme confines. The modeling studies applied to assist the design of analogs related to **10** were more involved. They included various cycles of energy minimizations with and without the enzyme binding site. The results led to better binding predictions and stronger binding in the resulting inhibitors, thus illustrating promising prospects when applying structure-based design in combination with computational methods. Yet, it was recognized that the ability to better quantify predictions was awaiting improved means to simulate the enzyme dynamics, to sample conformational population more exhaustively, and to include implicit and explicit solvation aspects in the calculations.

The concept of linking proximal side chains was also applied to the P2 and P4 side chains as seen in **11** and **12** (Table 3) [35, 36]. In the former case, Monte Carlo simulations of the cyclic tripeptide were carried out in the static active site, while in the latter example the inhibitor and the enzyme residues were both energy minimized. In an alternative approach, the cyclization of the P1 and P3 groups as illustrated by **13** was reported [37]. Clearly, the binding mode observed in the crystal structures inspired the cyclic inhibitor design, a direction that might not have been taken without the insight gleaned from the structural information.

The side-chain binding information led additionally to the successful design of inhibitors, which follow the lead of linking the side chains, but in an acyclic manner. 'Tethered' analogs, which linked the P1 and P3 groups but detached the P3 side chain from the backbone template (**14** and **15**, Table 4) are examples of this design [38]. Conformational SEARCH [39] analyses carried out with the 'tethered' side chains shown in **14** and **15**, supported the conformational validity of the modeled structures [40]. The potent activities for **14** and **15** further confirmed the continuum existing between the S1 and the S3 binding subsites (Fig. 6). However, attempts to truncate the backbone in the 'tethered' analogs by replacing Smo-Gly or Smo-Gly-Alg with an acetyl cap resulted in loss of activity (unpublished results). The removal of the P4-P3 backbone portion might reflect a minimum length requirement for the inhibitor backbone.

Table 4. Renin inhibitors that illustrate subsite interdependencies

IC_{50} (nM)

14.[a]   11

Table 4. (Continued)

	IC$_{50}$ (nM)

15.[a]

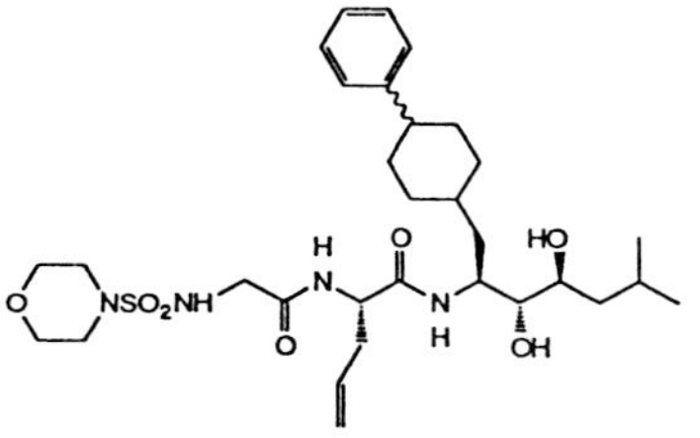

60.5

16.   BNMA[b]-NH ...

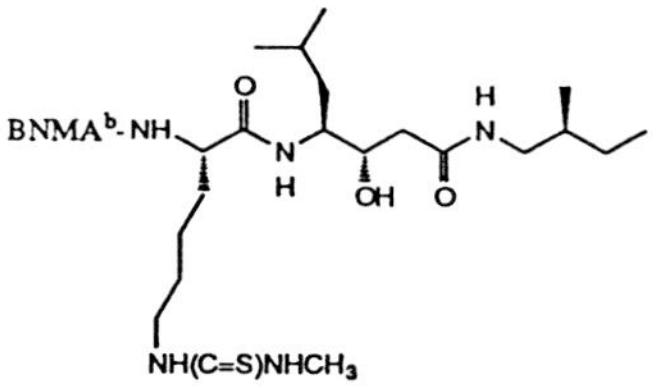

NH(C=S)NHCH$_3$

23

17.   BNMA-NH ...

NH(C=S)NHCH$_3$

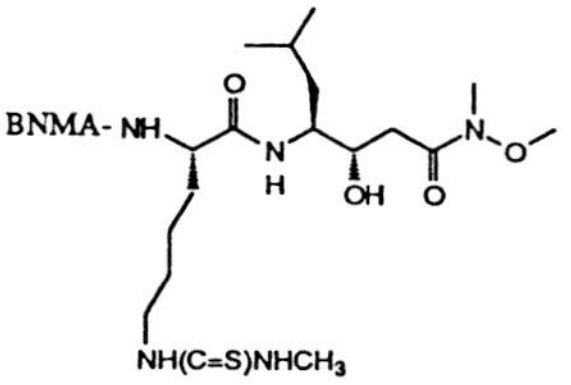

10,560

18.

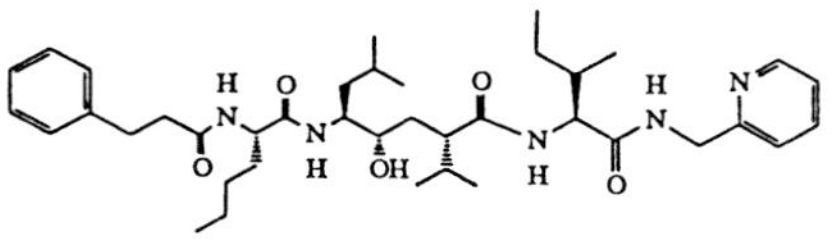

12

19.[c]

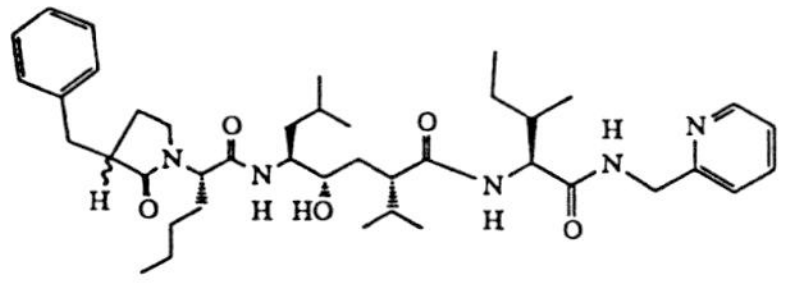

14
27

[a] Mixture of *cis/trans* isomers
[b] BNMA = bis-[(1-naphthyl)methyl]acetic acid
[c] Activities of separate epimers reported

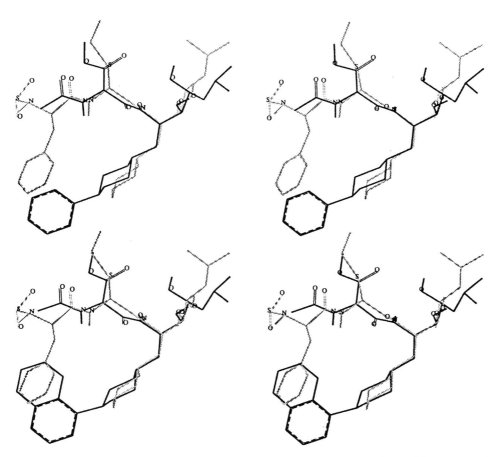

Figure 6. Stereoview of the bound conformation of a phenyl 'tethered' analog (upper, black lines) and a naphthyl 'tethered' analog (lower, black lines) overlaid with Smo-Phe-Ets-ACDMH (gray lines) extracted from the endothiapepsin X-ray crystal structure [17] and oriented in the renin model active site.

The concept of side chain interdependencies explained some unexpected structure-activity relationships. A series of potent inhibitors were designed which contained a flexible side chain at P2, capped with a thiourea group, as indicated by **16** (Table 4) [41]. Surprisingly, when the analog with an N,O-dimethylamide at P2′, **17**, was tested, the activity was significantly lower. Molecular modeling indicated that the *N,O*-dimethylamide group could bind in the S1′ pocket and might in fact compete for this binding site with the long flexible thiourea side chain. The analog retaining the *N,O*-dimethylamide group but replacing the long, flexible residue at P2 with His was designed with the intent of attenuating the subsite competition. The His analog did indeed retain potent binding (62 nM), supporting the hypothesis regarding subsite occupation.

Subsite interdependencies were not limited to adjacently binding groups as observed in the crystal structures [42]. In a series of constrained P3-P4 analogs, it was found that the potency was strongly dependent on the P1-P1′ isostere. As can be seen from Table 4, when a cyclic lactam was incorporated at the P4-P3 site of analogs containing a hydroxyethylene isostere at the P1-P1′ site, the potencies were equal to or, in cases not shown, better than the acyclic counterparts (**18** and **19**). However, when statine was inserted as the transition state mimetic, the compounds with the cyclic lactam were significantly less active than the acyclic analogs. This provided further evidence that the subsites were not independent entities and different binding modes occurred with different types of P1-P1′ groups. This dependency phenomenon was critical to consider in inhibitor design.

3.2.4 Renin Crystal Structure

The majority of the molecular modeling studies reported for renin involved a model of the enzyme. These structures were refined over the years as additional structural information became available and the tools to carry out the homology modeling improved. Purified human renin protein was extremely scarce and it was not until the later stages of the research that the enzyme crystal structure became available for use (Fig. 7) [43–46]. While overall, the renin structure resembles the topology of other aspartic proteases, the binding site is less open due in part to the relative location of a rigid body comprising the C-domain. Renin also contains a characteristic polyproline loop containing two *cis* amide bonds in the C-lobe. This

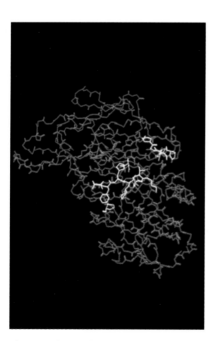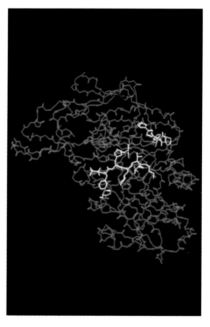

Figure 7. Stereoview of the human renin X-ray crystal structure bound with an inhibitor [43]. The inhibitor is shown in yellow and the enzyme in red with the catalytic aspartic acids, the polyproline loop and Pro111 on helix h_{N2} colored cyan.

polyproline segment, which includes Pro292-Pro294, and a loop involving residues 241–250, lie on either side of residues 72–81 in the flap region, in binding the inhibitors. Thus, loops from both lobes compose this residue lining, while with the other mammalian and fungal aspartic proteases, only the N-lobe is involved. Recently, an advanced renin model was compared with the experimentally determined structure of human renin [47]. The model was derived from the X-ray crystal structures of the aspartic proteases, pepsin and chymosin, using the program COMPOSER [48]. Superpositioning of the α-carbons indicated significant overlay of the model with the human renin X-ray structure. The major deviations existed at the surface-exposed loop regions, including the characteristic rigid polyproline loop, described above. This secondary structure is closer to the binding site in the renin X-ray complexes and forms in part the S1′ and S3′ pockets. Helix h_{N2}, which lines the S3 and S5 sites, is also oriented closer to the active site. While the comparison supports the usefulness of models derived from homologous enzymes, the structural differences also point out the ultimate desirability and advantage of obtaining an X-ray crystal structure for a more precise 3D representation of the enzyme.

3.2.5 Summary – Renin Modeling

The use of structure-based design in the discovery of human renin inhibitors was successful, but in a limited sense given the learning curve that was involved. Mainly through the use of an approximate human renin model, and manual docking procedures, large peptidic inhibitors were modified in a stepwise manner to remove a portion of their peptide nature and reduce their molecular weight. Unfortunately, the progress was slow, hampered by limitations that prevented dramatic alterations to produce a totally nonpeptide inhibitor. The obstacles included not only hardware and software resources available for use, but also the physical properties and binding requirements inherent with the target molecule.

Although molecular modeling could not quantitatively predict the potency of the proposed renin inhibitors, the inactivity of the noncompatible ligands could be more confidently forecast. As a result, considerable synthesis time was saved. Overall, structure-based design with renin provided the framework for ligand design and, perhaps more importantly, laid the foundation for the design efforts that are currently underway. It also provided an excellent training experience which led to the development of better computational tools and methods.

3.3 HIV-1 Protease

In the late 1980s, a new member of the aspartic protease family emerged as the target for inhibitor design, the HIV-1 protease. This macromolecule is found in the virus identified as the pathogen for the AIDS epidemic and is a member of a subclass of aspartic proteases found in retroviruses [49–51]. The HIV-1 protease processes polyproteins into structural and functional proteins and has been found to be essential for viral replication [52–54]. Before a 3D model for the HIV-1 protease was available, successful identification of inhibitors of the enzyme had been achieved. Applying the experience gained from the work with the renin enzyme, the 3D data available for various aspartic proteases and information known regard-

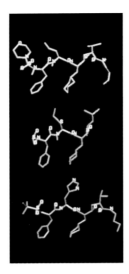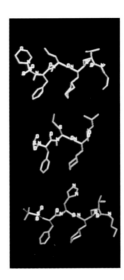

Figure 8. Stereoview of renin inhibitors extracted from the X-ray crystal structures of three different aspartic proteases. The yellow inhibitor, **20**, is from an HIV-1 protease structure (unpublished), the cyan inhibitor is from an endothiapepsin structure [17] and the green inhibitor is from a renin structure [43].

ing the HIV-1 protease and its substrates, a chemical databank of renin inhibitors was searched for inhibitors of the retroviral enzyme [55]. Through this process, two structurally related, potent inhibitors of HIV-1 protease, **20** and **21** (Table 5), were identified in an efficient and timely manner. In this application, structure-based knowledge takes on a somewhat different role in advancing inhibitor discovery, in the identification of active ligands from existing compounds. This indeed can provide a tremendous headstart in inhibitor design. Eventually, an X-ray crystal structure of each inhibitor bound to HIV-1 protease was solved (to be published). The bound inhibitor, **20**, was compared to structurally related renin inhibitors cocrystallized with human renin and the fungal protease, endothiapepsin. The similarity of the binding modes across the subclasses of the aspartic proteases was apparent (Fig. 8). The results indicated that although the overall sequence homology across the subclasses of aspartic proteases is not high, the binding modes of the inhibitors are conserved to a significant degree. Thus, the active site regions would be expected to be considerably more structurally similar than the sequence alignment would suggest.

3.3.1 3D Structures of HIV-1 Protease

Prior to the determination of the X-ray crystal structure of HIV-1 protease, a model of the enzyme was derived from the homologous RSV protease [56]. Within a short amount of time, the first crystal structures of the native HIV-1 protease were solved [57, 58], followed by numerous enzyme/inhibitor crystal complexes [59]. These experimental results not only confirmed the accuracy of the model with regard to the substrate binding residues, but provided

Table 5. HIV-1 protease inhibitors

IC$_{50}$ (nM)

	IC$_{50}$ (nM)
20.	15
21.[a]	2
22.	100,000[b]
23.	15,000[b]
24.[c]	1[b]

[a] Mixture of diastereomers
[b] K_i (nM)
[c] Stereochemistry not reported

Table 5   (Continued)

	IC$_{50}$ (nM)
25.	0.60[b]
26.	9,000
27.	0.27[b]
28.	10,000[b]

an experimentally derived structure for molecular modeling studies at the initial stages of inhibitor design (Fig. 9). This was in sharp contrast to the availability of X-ray structures in the renin endeavor, and represented a significant advantage for structure-based design in the HIV-1 protease research.

The HIV-1 protease, as a member of the aspartic protease family of enzymes, contains an aspartic acid dyad at the catalytic site. The enzyme is active as a homodimer, with each subunit containing 99 amino acids (the numbering system is 1–99 for the first monomers and 101–199 for the second). This differs from the monomeric structure of renin, which is composed of 340 residues. The identical monomers in HIV-1 protease produce a symmetric

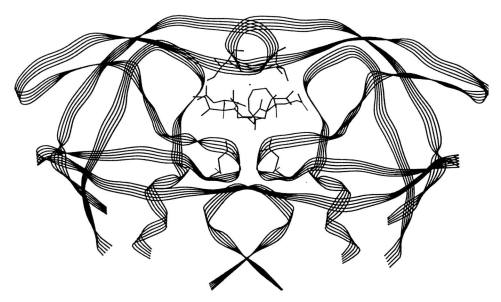

Figure 9. Ribbon representation of the HIV-1 protease X-ray crystal structure bound with an inhibitor [58].

enzyme having two flexible flap regions, one from each subunit, that cover the bound ligand. Interestingly, as described above, the renin crystal structure revealed that its binding site is covered by flaps from both lobes, instead of only the N-lobe as is seen in other mammalian and fungal aspartic proteases. This results in a superficial resemblance to the two equivalent flap regions found in the retroviral proteases [44]. With regard to the ligand binding, the peptidomimetic inhibitors bind to HIV-1 protease in an extended conformation with the transition state mimetics oriented at the catalytic dyad. This is analogous to that observed for inhibitors of renin and other mammalian as well as fungal aspartic proteases. However, unlike these inhibitors, the retroviral HIV-1 protease binds a conserved water, H_2O301, between the flap region and the ligand backbone. As we shall discuss, this binding phenomenon that was elucidated by structure-based design makes possible a novel, nonpeptide inhibitor design strategy that proved to be extremely successful for HIV-1 protease but nonexistent for renin.

3.3.2 HIV-1 Protease Nonpeptide Inhibitors

In recent years a number of potent peptidomimetic inhibitors of HIV-1 protease have been reported in the literature. One highly potent series was designed to be symmetric molecules to complement the enzyme [60] (Fig. 10). However, as with renin inhibitors, these analogs often carry with them the properties associated with peptides: low cellular activity and poor oral bioavailability. The quest, therefore, for a nonpeptide inhibitor again became the focus of our efforts as well as that of numerous other laboratories. In contrast to the renin studies,

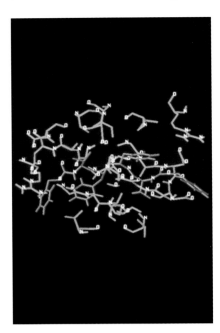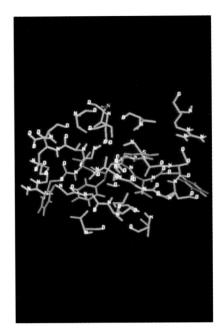

Figure 10. Stereoview of the X-ray crystal structure of the symmetric inhibitor A-74704 (magenta) bound in the HIV-1 protease (cleft region, cyan) [82]. H$_2$O301 is shown in green.

totally nonpeptide leads were determined early in the research through the use of diverse methods and technologies, not available even during the peak of renin research. In one of the first and most innovative endeavors, the use of the program DOCK [61] to evaluate the fit of rigid 3D structures from the Cambridge Crystallographic Database [31] in the HIV-1 protease cleft was reported [62]. The binding site was represented by a series of nonconcentric spheres. This led to the identification of haloperidol, **22**, as a weak inhibitor, but nonetheless an initial lead that was elaborated to more potent inhibitors including the thioketal derivative, **23** (Table 5). Although in the original X-ray structure of **23** with the protease, the inhibitor did not bind in the manner predicted by DOCK, a second crystal structure using Q7K HIV-1 protease revealed two binding modes for the inhibitor including one similar to that found with DOCK [63]. The information gleaned from this study points to the possibility of multiple binding modes for small inhibitors as well as the simultaneous binding of two ligands. To ensure a more accurate computational evaluation, the authors found that inclusion of conformational flexibility in the ligand was essential.

The application of *de novo* design using HIV-1 protease crystal structures led to the discovery of very potent inhibitors [64]. This iterative process of modeling, synthesis, and multiple X-ray crystal structures elaborated an initial lead having an IC$_{50}$ of 60 μM to the potent nanomolar inhibitor, **24**, possessing significant clinical potential. An iterative *de novo* inhibitor design was also applied in the discovery of a second series of extremely potent, bioavailable inhibitors, including **25** [65]. A third example of *de novo* design [66] was unique in that

Table 6. Coumarin and pyrone inhibitors of HIV-1 protease

IC$_{50}$ (nM)

		IC$_{50}$ (nM)
29.		2,300
30.		3,000
31.		1,670
32.		5,500
33.[a]		58
34.[a]		500[b]

[a] Enantiomeric mixture
[b] K_i (nM)
[c] Mixture of four stereoisomers

Table 6   (Continued)

IC$_{50}$ (nM)

35.[c]	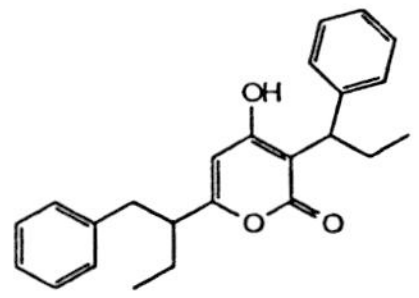 38[b]
36.[a]	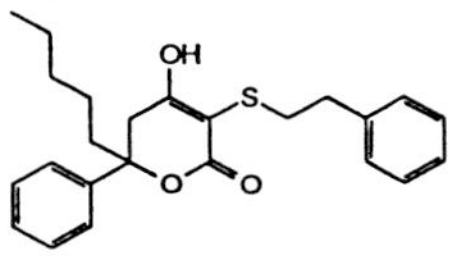 84

the ligand was designed to bind at the catalytic site and simultaneously displace H$_2$O301 in the flap region, thus, functioning as a hydrogen bond acceptor directly with the protease. These studies resulted in six-membered ring analogs represented by **26** in Table 5.

In another study, a 3D database search using a pharmacophore model that contained the intramolecular distances among the P1 and P1′ groups and the donor/acceptor bound at the catalytic site was carried out [67]. A 'hit' was found, which besides meeting the pharmacophore criteria, also presented an oxygen that matched the H$_2$O301 location. This initial lead was then elaborated into seven-membered ring analogs that incorporated a urea functionality. Compounds in this series, for example **27**, were potent, selective inhibitors of the protease (Table 5). They retained the C2 symmetry found in the peptidomimetic design, displaced H$_2$O301 upon binding, and possessed relatively low molecular weight and high oral bioavailability. In a third study involving H$_2$O301 displacement, a modified DOCK program [68] was used to identify structures in a company database that would bind at the catalytic site and simultaneously displace H$_2$O301 in the active site of the protease [69]. The resulting 'hits' suggested six-membered ring analogs with 1–4 related oxygens, as illustrated by **28**. Although only moderate activity was obtained with this analog, modifying the structure with hydrophobic groups that could bind in the S2 and S2′ pockets might significantly enhance the potency.

Simultaneously, the advanced technology of mass screening was applied to an extensive chemical database [70]. This process resulted in numerous 'hits' including the coumarin and pyrone, **29** and **30**, shown in Table 6. Both inhibitors exhibited micromolar potency. Because of their small size, achiral nature and their inhibitory activity, they became the prototypes for biological and chemical studies as well as for X-ray determination. The mass screening approach was also applied elsewhere and resulted in a very similar warfarin 'hit' [71]. In both cases, the initial micromolar leads were elaborated to potent inhibitors through structure-based design involving the collaboration of molecular modeling, synthesis and X-ray crystallography.

3.3.3 Docking/Modeling HIV-1 Protease Nonpeptide Inhibitors

While the crystallographic studies were initiated for the mass screening 'hit', **29**, and an analog related to the pyrone 'hit', **31**, molecular modeling was undertaken to determine the bound conformations of these nonpeptides [72]. This presented a new challenge in structure-based design, since the modeling of the renin inhibitors generally involved ligand structures, which retained at least to some degree the peptide-like scaffold. Although the X-ray information available for the HIV-1 protease lent a great deal of guidance in the docking of peptide-like inhibitors, various questions were raised regarding the docking of the small molecules that were completely nonpeptide-like. The variables considered as the modeling studies were initiated included the orientation of the flexible hairpin loops, the retention of conserved water molecules upon binding, the possibility of multiple binding modes, and a greater than 1/1 stoichiometry of binding for the ligand/protease complex. The program AUTODOCK [73] was used in docking the nonpeptide analogs. This program applies Monte Carlo simulations with simulated annealing to find the most favorable binding orientation and conformation for a ligand in a rigid receptor. The enzyme structure for the protease/MVT101 complex was used in the docking procedure; thus, the flap regions were positioned in the closed orientation. Although the original X-ray with the nonpeptide haloperidol derivative showed the flaps only partially closed, it was reasoned that the closed orientation would most likely afford tighter binding for the complex. Even though the peptidomimetic inhibitors bind with a conserved water molecule at the flap region, it could be postulated that these nonpeptide analogs displace H_2O301. Conversely, the novel inhibitors might not bind as transition state mimetics and thus, not displace the catalytic water found in the active site of the native enzyme. To more comprehensively cover the possibilities regarding the conserved bound water molecules, docking experiments for **29** were carried out with three different solvation states of the enzyme: the first state would retain both the flap water, H_2O301, and the catalytic water (H_2OCAT), the second only H_2O301, and the third no solvent molecules. The lowest energy results from the simulations carried out at each solvation state were analyzed and indicated binding centered between the S2 and S2′ sites. The most interesting results were found for the simulation run with no water. This result showed that the coumarin system could span the width of the binding cavity and interact simultaneously with the catalytic dyad and Ile50/Ile150 in the flap region (Fig. 11).

The coumarin X-ray crystal structure was eventually solved and showed, interestingly, that the inhibitor could actually bind in two distinct modes, each of which displaced both conserved water molecules (Fig. 12). The first binding mode was clearly predicted from the docking studies carried out with no water (Fig. 13). The second mode shifted the inhibitor to the prime binding region positioning the flexible side chain in the S2′ site. The hydroxyl group could interact with each of the oxygens of the dyad carboxylate groups but the lactone only interacted with Ile50(NH). These results indicated that small inhibitors can undertake alternative binding modes and retain comparable binding affinity. The elucidation of the binding modes could now be applied to inhibitor design.

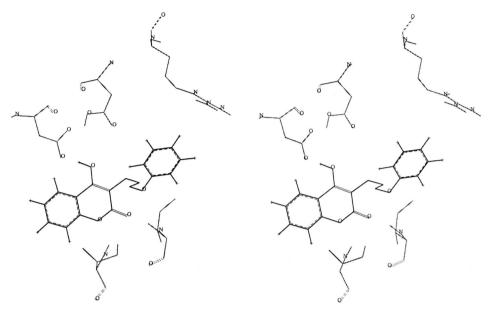

Figure 11. Stereoview of the bound conformation of **29** (black lines) resulting from an AUTODOCK simulation with HIV-1 protease [72]. Asp25/Asp125 (the catalytic aspartic acids), Ile50/Ile150 and Arg108 are also shown (gray lines).

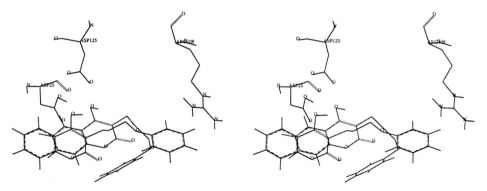

Figure 12. Stereoview of the two binding modes found for **29** in the X-ray crystal structure with HIV-1 protease. The enzyme residues, Asp25/Asp125 and Arg108, are from the structure bound with the inhibitor conformer represented by the black lines.

Since the coumarin structure occupied fewer pockets than the peptidomimetic analogs (Fig. 14) and the two binding modes in the crystal structure showed that the coumarin side chain could occupy two separate binding sites, the concept of branching the phenoxy side chain was pursued. This resulted in **32**, which incorporated a nitrogen at the branching point

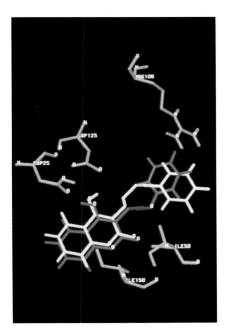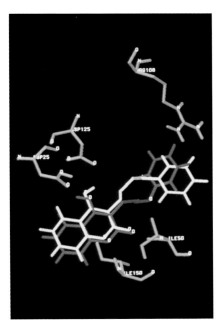

Figure 13. Stereoview of an overlay of the bound conformation of **29** (red) resulting from an AUTO-DOCK simulation with one binding mode of **29** (yellow) found in the X-ray crystal structure with HIV-1 protease [72]. The enzyme residues Asp25/Asp125, Ile50/150 and Arg108 from the X-ray structure are shown in cyan.

to retain achirality. A small drop in potency was observed, but the strategy of branching proved feasible. The slightly reduced activity could be linked to the lack of the oxygen in the side chain. It was found that replacement of the oxygen in the parent compound with a sulfur or a methylene group attenuated the activity by 10- and 50-fold, respectively. Surprisingly, when both binding modes of **29** in the crystal structure were analyzed, no direct polar contact with the protease cleft was found. Since the X-ray structure was not refined to a resolution that would determine the position of water molecules, the possibility of a water mediated interaction could not be ruled out. To investigate this further, the program GRID [74] was employed. Through GRID calculations, favorable interaction sites for atomic and molecular probes can be determined for a target molecule. Using a water probe and the inhibitor/protease complex, the low energy regions for bound water molecules could be identified and displayed as energy contours. As can be seen in Fig. 15 (a and b), for both of the coumarin binding modes, a favored interaction site for a water molecule can be found between the side chain oxygen and the enzyme. In the first mode of binding the mediated interaction involves residue Asp29 and in the second mode, Gly48 in the flap region engages in the contact. Therefore, the importance of a hydrogen bond acceptor in the side chain to interact with a water molecule could explain the drop in activity observed with the analogs having the sulfur and methylene replacements, as well as the branched nitrogen inhibitor, **32**.

Figure 14. Stereoview of an overlay of the HIV-1 protease bound conformations of **29** (magenta) and A-74704 (white) extracted from the X-ray crystal structures [72, 82]. H2O301 is shown in green.

In the pyrone series [75], developed from the second mass screening 'hit', docking studies were carried out for **30** and the related analog **31**, using molecular dynamics simulations and AUTODOCK [76]. Both applications were found to be predictive of the binding mode found in the subsequently determined crystal structure of **31**; however, the X-ray complex served to rule out alternative binding modes that could not have been disqualified based simply on the docking experiments. In the case of the AUTODOCK calculations for **31** involving no water molecules, two modes of binding seemed feasible (to be published). In both cases, the hydroxyl group bound at the catalytic dyad and the lactone carbonyl bound to the flap region. However, in one mode the phenyl and benzyl groups occupied the P1 and P1′ pockets, whereas in the second orientation they bound in the P2 and P2′ sites. When the crystal structure was determined, the former mode of binding was found to exist (Fig. 16).

In studying the binding of these inhibitors, the program CAVITY [77] was applied with the HIV-1 protease binding site. Using a flood-fill algorithm, this program ultimately produces a graphical display, which affords insightful visualization of the size, shape and polarity of the enzyme binding site, as is illustrated in Fig. 17 for HIV-1 protease.

The strategy of branching to reached unoccupied pockets was also applied in the pyrone series [78]. As can be seen by **33**, a significant increase in potency was observed (Table 6). Interestingly, the X-ray crystal structure of **33** showed a shift in the orientation of the inhibitor in the active site relative to the parent achiral inhibitor, **31** (Fig. 18). The hydroxyl group no longer interacted with both catalytic aspartates, but bound solely to Asp125. However, a

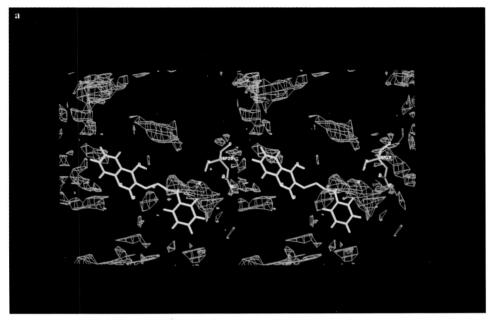

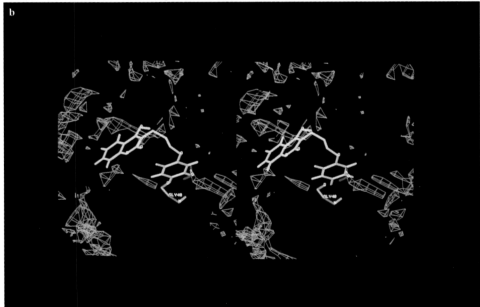

Figure 15. (a) Stereoview of the contour at −5 kcal/mol for the GRID calculation using a water probe (cyan) and the HIV-1 protease crystal structure [72]. Inhibitor **29** (yellow, first binding mode), an inserted water molecule (red), and Asp29 (magenta) are shown. (b) Stereoview of the contour at −5 kcal/mol for the GRID calculation using a water probe (cyan) and the HIV-1 protease crystal structure [72]. Inhibitor **29** (yellow, second binding mode), an inserted water molecule (red), and Gly48 (magenta) are shown.

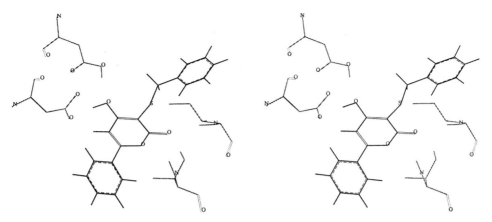

Figure 16. Stereoview of **31** (black lines) bound in the HIV-1 protease crystal structure [75]. Asp25/Asp125 and Ile50/Ile150 are displayed (gray lines).

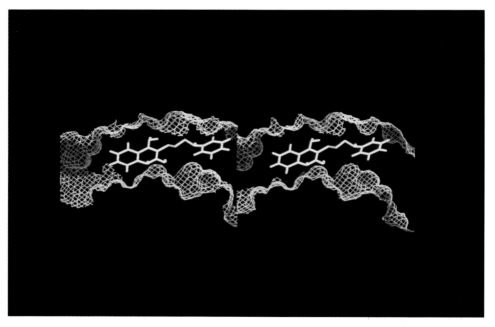

Figure 17. Stereoview of a cast of the HIV-1 protease binding site in the X-ray crystal structure [72] generated using CAVITY [77]. The CAVITY display is color-coded by electrostatics. Analog **29** is shown in yellow.

water molecule mediated an interaction between the pyrone hydroxyl and Asp25. In work involving similar compounds, the protease X-ray crystal structure with a branched pyrone inhibitor, **34**, was reported [71]. The mode of binding for this inhibitor was found to be similar

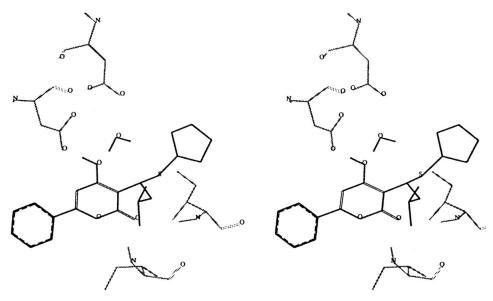

Figure 18. Stereoview of **33** and the bridging water (black lines) bound in the HIV-1 protease crystal structure [78]. Asp25/Asp125 and Ile50/Ile150 (gray lines) are also shown.

to that observed for **31** bound to the protease. These results suggest that the groups making up the branched system dictate the preferred binding mode. They also raise the question regarding the importance of the sulfur in binding the mediating water molecule, which is located 3.4 Å away from the sulfur atom.

The protease X-ray crystal structure of **34** showed that the α-ethyl and the α-phenyl groups were bound in the S1 and S2 sites, respectively, and that the phenethyl group was positioned near the S2′ pocket. Compound **35** was then designed with the aim of gaining additional binding by placing a second ethyl group in the S1′ site [71]. This compound was tested as a mixture of four stereoisomers and exhibited a 13-fold increase in potency relative to **34**.

A second approach to filling additional pockets in the protease binding site with this nonpeptide series was pursued. Molecular modeling indicated that a dihydropyrone system, in place of the pyrone ring, could also displace H_2O301 and bind a 4-hydroxyl group to the catalytic dyad [79]. More significantly, branching at position 6 of the dihydropyrone ring was compatible with the protease binding site. This prediction was confirmed when a series of 6,6-disubstituted dihydropyrone analogs, including **36**, were found to be potent inhibitors of HIV-1 protease (compare **36** with **31**). Compound **36** was co-crystallized with the HIV-1 protease as a racemic mixture. Unfortunately, the electron density observed for the ligand was not defined well enough to distinguish which enantiomer was bound or indeed, whether both enantiomers were bound (Fig. 19). This result pointed to the importance of having enantiomerically pure compounds in this series for the X-ray crystallographic studies.

The two binding modes observed for the pyrone series, the dual binding orientations found for the single coumarin inhibitor, and the stoichiometry of binding revealed for the haloper-

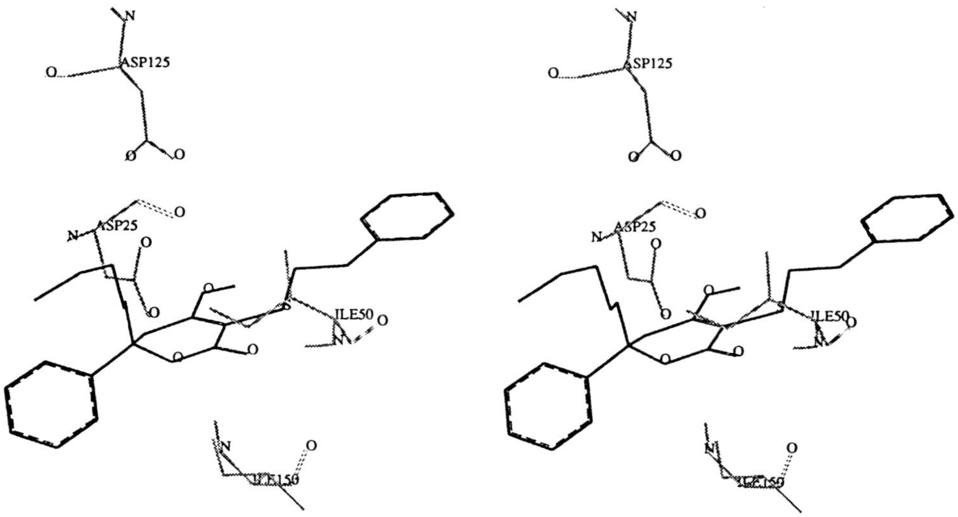

Figure 19. Stereoview of **36** (black lines) bound in the HIV-1 protease crystal structure (the S enantiomer is fitted to the electron density) [79]. Asp25/Asp125 and Ile50/Ile150 (gray lines) are displayed.

idol derivative highlight the underlying complexity of structure-based design for small, nonpeptide inhibitor design. We have not only found significantly different binding modes for related analogs, but multiple modes of binding for a single compound. These results stress the importance of multiple inhibitor/enzyme X-ray crystal structures to more comprehensively understand the binding interactions. The newly acquired information increases the potential for successful novel inhibitor design.

3.4 Summary: Comparison of HIV-1 Protease versus Renin Structure-Based Design

An evaluation of stucture-based design, as it was applied with the two aspartic proteases, HIV-1 protease and renin, serves as an excellent gauge of how the approach has progressed during the past decade. It highlights the various technological advancements that have elevated the method to higher levels and it points to inherent structural and physical features of the enzymes that impact the ultimate potential of structure-based design for each protease.

Technological breakthroughs in the past decade have made possible more sophisticated techniques that could be applied to the HIV-1 protease research. 3D database searching, along with the state-of-the-art chemical mass screening have helped make possible the identification of totally nonpeptide, low-molecular weight inhibitor leads. These types of advantageous starting points have been a genuine boost for structure-based design that were not realized during the studies with renin.

There has literally been an explosion in the emergence of molecular modeling and graphics software which, together with the enhanced hardware capabilities, have allowed computational feats unheard of ten years ago. Advanced software programs, e.g., AUTODOCK, GRID and CAVITY aid in the elucidation of ligand binding and move us forward in design strategies. With HIV-1 protease, these methodologies could be applied to 3D crystal coordinates of the enzyme in the initial stages of the studies. Within a few years, X-ray crystallographic techniques made possible the generation of numerous complexes. These structures provided critical information regarding binding, particularly for the small nonpeptides that can adopt altered binding modes as they are modified. This was clearly a significant advantage for the HIV-1 protease work versus the renin studies. As we have noted, the renin crystal structure became available only in the last stages of the research and to this date only a handful of complexes have been reported.

With regard to the ligand binding modes, small nonpeptide inhibitors of HIV-1 protease can take advantage of displacing the uniquely conserved H_2O301 bound at the flap region. The favorable entropic effect gained upon binding directly to the enzyme may allow for low-molecular weight ligands to bind with high affinity and simultaneously excellent selectivity versus the mammalian and fungal enzymes. As was pointed out, renin has no analogous binding phenomenon on which to capitalize.

The capability to communicate 3D structural studies with the chemists and other researchers has evolved significantly [80]. Today in our laboratories, the use of a 3D projection room has advanced the HIV-1 protease inhibitor design to a team effort, in which creative design ideas can be discussed within a group in an extremely productive manner, while visualizing the protein/inhibitor interactions.

3.5 Current Limitations/Future Perspective

One can speculate how renin research would fare today. Certainly the multiple software packages that have appeared in recent years along with the increases in computer performance would result in more elegant modeling and greater insight regarding binding, as was observed with HIV-1 protease. However, one has to wonder if there are features inherent with renin that would prevent it from realizing the success that has been achieved for HIV-1 protease structure-based design. Would the problem of inadequate protein supply be an obstacle that could be surmounted today? And if it were, would the protein be found to crystallize as readily as HIV-1 protease? Renin is very selective for angiotensinogen, whereas HIV-1 protease is promiscuous. Is this feature indicative that a potent ligand for renin would need to remain at least to some degree peptide-like and relatively large? These questions point out that although we have advanced by 'leaps and bounds' in the technology that we can apply to structure-based design, the targeted protein can provide limitations of its own. This is particularly true for those cases in which the natural ligand is a peptide or protein and the design strategy is aimed at drug discovery, that is, an orally active agent.

In receptor structure-based design the operative word is structure. The success of the methodology is closely tied to the amount of structural information that is available. Although a receptor model derived from experimentally determined structures of homologous proteins is adequate, as was the case with renin, having an experimentally determined structure of the targeted macromolecule is desirable. Having multiple complexes would be the 'ideal situation' as the work with HIV-1 protease has illustrated. However, not all proteins are readily cloned, isolated or crystallized. Once the advancements in these areas can provide the 'ideal situation' on a standard basis, there is no doubt structure-based design will become a norm in the drug discovery field.

Although the software and hardware tools used in structure-based design have progressed, limitations continue to exist in the area of molecular modeling. These include the ability to adequately handle dynamics, solvation, and free energy calculations. Advances in computational performance improve the capability of dealing with these weaknesses. A recent review pointed out that related limitations exist in the current docking methods, while forecasting that automated docking will play an ever-important role in discovering lead compounds [81]. In the area of computer graphics and hardware, work is ongoing in the development of faster algorithms to produce higher quality images to represent, for example, electrostatic potential, molecular orbitals and electron density [80]. At the same time the use of parallel supercomputers increases the speed with which applications can be carried out and also makes calculations on large systems feasible.

3.6 Conclusion

Receptor structure-based design is a powerful tool that can direct ligand design in a rational, efficient and inspired manner. The potential of the approach is closely linked to the availability of structural information and the associated hardware and software tools. In the past 10 years, we have witnessed the tremendous technological advances that have taken place in these areas and have in turn observed the progress made in structure-based design. Considering the pace with which the technology continues to move forward, the prospect for increased success with the structure-based design approach is undeniable. Concurrent with this approach is the accumulation of experience and knowledge that impacts the future applications and directions. Regardless of the limitations confronting the renin research, we were able to study and evaluate inhibitor/enzyme interactions, define the constraints of the binding site and analyze the more subtle aspects of binding such as subsite interdependencies. We could develop a better understanding and appreciation of what enhances receptor recognition and ligand binding, and what counteracts them. Clearly, the renin research served as a learning process that formed the groundwork for what we experience today with HIV-1 protease and other structure-based design endeavors. In the HIV-1 protease work, we have seen the successful design of totally nonpeptide, orally active inhibitors through the use of multiple X-ray structures of inhibitor/enzyme complexes. We have witnessed the unique binding mechanism displayed by nonpeptide inhibitors that involves displacement of a water molecule bound in the flap region. Through the use of advanced software we are able to predict binding modes and map out interaction sites. What we have learned with HIV-1 protease and other current applications in turn adds to the wealth of knowledge in structure-based design and plays a critical role in moving the methodology forward. It is evident that structure-based design will be with us for some time to come. Its future looks extremely bright and its true potential is far from being realized.

References

[1] Verlinde, C. L., and Hol, W. G., *Structure* **2**, 577–587 (1994)

[2] Greer, J., Erickson, J. W., Baldwin, J. J., and Varney, M. D., *J. Med. Chem.* **37**, 1035–1054 (1994)

[3] Reich, S. H., and Webber, S, E., *Perspect. Drug Discovery Des.* **1**, 371–390 (1993)

[4] Kunst, I. D., *Science* **257**, 1078–1082 (1992)

[5] Navia, M. A., and Murcko, M. A., *Curr. Opin. Struct. Biol.* **2**, 202–209 (1992)

[6] Whittle, P. J., and Blundell, T. L., *Annu. Rev. Biomol. Struct.* **23**, 349–375 (1994)

[7] Greenlee, W. J., *Medicinal Research Reviews* **10**, 173–236 (1990)

[8] Bott, R., Subramanian E., and Davies, D. R., *Biochemistry* **21**, 6956–6962 (1982)

[9] Pearl, L. H., and Blundell, T. L., *FEBS Lett.* **174**, 96–101 (1984)

[10] James, M. N. G., and Sielecki, A., *J. Mol. Biol.* **163**, 299–301 (1983)

[11] Carlson, W., Karplus M., and Haber, E., *Hypertension* **7**, 13–26 (1985)

[12] Plattner, J. J., Greer, J., Fung, A. K., Stein, H., Kleinert, H. D., Sham, H. L., Smital, J. R., and Perun, T. J., *Biochem. Biophys. Res. Commun.* **139**, 982–990 (1986)

[13] Hutchins, C., and Greer, J., *Crit. Rev. Biochem. Mol. Biol.* **26**, 77–127 (1991)

[14] Sibanda, B. L., Blundell, T., Hobart, P. M., Fogliano, M., Bindra, J. S., Dominy, B. W., and Chirgwin, J. M., *FEBS Lett.* **174**, 102–111 (1984)

[15] Blundell, T. L., Jenkins, J. A., Sewell, B. T., Pearl, L. H., Cooper, J. B., Wood, S. P., Veerapandian, B., *J. Mol. Biol.* **211**, 919–941 (1990)

[16] Suguna, K., Padian, E. A., Smith, C. W., Carlson, W. D., Davies, D. R., *Proc. Natl Acad. Sci., USA* **84**, 7009–7013 (1987)

[17] Lunney, E. A., Hamilton, H. W., Hodges, J. C., Kaltenbronn, J. S., Repine, J. T., Badasso, M., Cooper, J. B., Dealwis, C., Wallace, B. A., Lowther, W. T., Dunn B. M., and Humblet, C., *J. Med. Chem.* **36**, 3809–3820 (1993)

[18] Sali, A., Veerapandian, B., Cooper, J. B., Moss, D. S., Hofmann, T., and Blundell, T. L., *Proteins: Struct., Funct. Genet.* **12**, 158–170 (1992)

[19] Schechter, I., and Berger, A. The 'P' subsite nomenclature relates amino acid residues or mimics in the inhibitor to corresponding residues in the natural substrate angiotensinogen. The 'S' nomenclature relates in terms of the enzyme subsites. *Biochem. Biophys. Res. Commun.* **27**, 157–162 (1967)

[20] Verhoef, J. C., Bodde, H. E., de Boer, A. G., Bouwstra, J. A., Junginger, H. E., Merkus, F. W., and Breimer, D. D., *Eur. J. Drug Metab. Pharmacokinet.* **15**, 83–93 (1990)

[21] Boger, J., Renin Inhibitors. Design of angiotensin transition-state analogs containing statine. In: *Peptides, Structure and Function. Proceedings of the Eighth American Peptide Symposium,* Hruby, V. J., and Rich, D. H. (Eds.). Pierce Chemical Co.: Rockford, IL; 569–578 (1983)

[22] Boger, J., Payne, L. S., Perlow, D. S., Lohr, N. S., Poe, M., Blaine, E. H., Ulm, E. H., Schorn, T. W., LaMont, B. I., Lin, T-Y., Kawai, M., Rich, D. H., and Veber, D. F., *J. Med. Chem.* **28**, 1779–1790 (1985)

[23] Hanson, G. J., Baran, J. S., Lowrie, H. S., Russell, M. A., Sarussi, S. J., Williams, K., Babler, M., Bittner, S. E., Papaioannou, S. E., Yang, P-C., and Walsh, G. M., *Biochem. Biophys. Res. Commun.* **160**, 1–5 (1989)

[24] Hanson, G. J., Baran, J. S., Clare, M., Williams, K., Babler, M., Bittner, S. E., Russell, M. A., Papaioannou, S. E., Yang, P.-C., and Walsh, G. M. Orally active renin inhibitors containing a novel aminoglycol dipeptide (Leu-Val) mimetic. In: *Peptides, Chemistry, Structure and Biology, Proceedings of the Eleventh American Peptide Symposium,* Rivier, J. E., and Marshall, G. R., (Eds.). ESCOM: Leiden, Netherlands; 396–398 (1989)

[25] Luly, J. R., BaMaung, N., Soderquist, J., Fung, A. K. L., Stein, H., Kleinert, H. D., Marcotte, P. A., Egan, D. A., Bopp, B., Merits, I., Bolis, G., Greer, J., Perun, T. J., and Plattner, J. J., *J. Med. Chem.* **31**, 2264–2276 (1988)

[26] Rosenberg, S. H., Dellaria, J. F., Kempf, D. J., Hutchins, C. W., Woods, K. W., Maki, R. G., de Lara, E., Spina, K. P., Stein, H. H., Cohen, J., Baker, W. R., Plattner, J. J., Kleinert, H. D., and Perun, T. J., *J. Med. Chem.* **33**, 1582–1590 (1990)

[27] Repine, J. T., Himmelsbach, R. J., Hodges, J. C., Kaltenbronn, J. S., Sircar, I., Skeean, R. W., Brennan, S. T., Hurley, T. R., Lunney, E., Humblet, C. C., Weishaar, R. E., Rapundalo, S., Ryan, M. J., Taylor, D. G., Olson, S. C., Michniewicz, B. M., Kornberg, B. E., Belmont, D. T., and Taylor, M. D., *J. Med. Chem.* **34**, 1935–1943 (1991)

[28] Lunney, E. A., and Humblet, C. C. Comparative molecular modeling analyses of endothiapepsin complexes as renin model templates. In: *Peptides: Chemistry, Structure and Biology, Proceedings of the Eleventh American Peptide Symposium,* Marshall, G. R., and Rivier, J. E. (Eds.). ESCOM: Leiden, Netherlands; 387–389 (1990)

[29] Cooper, J., Quail, W., Frazao, C., Foundling, S. I., Blundell, T. L., Humblet, C., Lunney, E. A., Lowther, W. T., Dunn, B. M., *Biochemistry* **31**, 8142–8150 (1992)

[30] de Laszlo, S. E., Bush, B. L., Doyle, J. J., Greenlee, W. J., Hangauer, D. G., Halgren, T. A., Lynch, R. J., Schorn, T. W., and Siegl, P. K. S., *J. Med. Chem.* **35**, 833–846 (1992)

[31] (a) Allen, F. H., Kennard, O., and Taylor, R., *Acc. Chem. Res.* **16**, 146–153 (1983); (b) Allen, F. H., Bellard, S., Brice, M. D., Cartwright, B. A., Doubleday, A., Higgs, H., Hummelink, T., Hummelink-Peters, B. G., Kennard, O., Motherwell, W. D. S., Rodgers, J. R., and Watson, D. G., *Acta Crystallogr.* **B35**, 2331–2339 (1979)

[32] Reily, M. D., Thanabal, V., Lunney, E. A., Repine, J. T., Humblet, C. C., and Wagner, G., *FEBS Lett.* **302**, 97–103 (1992)

[33] Weber, A. E., Halgren, T. A., Doyle, J. J., Lynch, R. J., Siegl, P. K. S., Parsons, W. H., Greenlee, W. J., and Patchett, A. A., *J. Med. Chem.* **34**, 2692–2701 (1991)

[34] Weber, A. E., Steiner, M. G, Krieter, P. A., Colletti, A. E., Tata, J. R., Halgren, T. A., Ball, R. G., Doyle, J. J., Schorn, T. W., Stearns, R. A., Miller, R. R., Siegl, P. K. S., Greenlee, W. J., and Patchett, A. A., *J. Med. Chem.* **35**, 3755–3773 (1992)

[35] Thaisrivongs, S., Blinn, J. R., Pals, D. T., and Turner, S. R., *J. Med. Chem.* **34**, 1276–1282 (1991)

[36] Greer, J., Hutchins, C., Bolis, G., Fung, A., and Sham, H. Comparative modeling of proteins in the design of novel cyclic renin inhibitors. In: *Proteins,* Renugoplakrishnan, V. (Ed.). ESCOM: Leiden, Netherlands; 283–289 (1991)

[37] Brotherton-Pleiss, C. E., Newman, S. R., Waterbury, L. D., and Schwartzberg, M. S. Design and synthesis of conformationally restricted renin inhibitors. In: *Peptides: Chemistry and Biology, Proceedings of the Twelfth American Peptide Symposium,* Smith, J. A., and Rivier, J. E. (Eds.). ESCOM: Leiden, Netherlands; 816–817 (1992)

[38] Plummer, M., Hamby, J. M., Hingorani, G., Batley, B. L., and Rapundalo, S. T., *Bioorg. Med. Chem. Lett.* **3**, 2119–2124 (1993)

[39] Motoc, I., Dammkoehler, R. A., Mayer, D., and Labanowski, J., *Quant. Struct.-Act. Relat.* **5**, 99–105 (1986)

[40] Plummer, M. S., Shahripour, A., Kaltenbronn, J. S., Lunney, E. A., Steinbaugh, B. A., Hamby, J. M., Hamilton, H. W., Sawyer, T. K., Humblet, C., Doherty, A. M., Taylor, M. D., Hingorani, G., Batley, B. L., and Rapundalo, S. T., *J. Med. Chem.* **38**, 2893–2905 (1995)

[41] Doherty, A. M., Kaltenbronn, J. S., Hudspeth, J. P., Repine, J. T., Roark, W. H., Sircar, I., Tinney, F. J., Connolly, C. J., Hodges, J. C., Taylor, M. D., Batley, B. L., Ryan, M. J., Essenburg, A. D., Rapundalo, S. T., Weishaar, R. E., Humblet, C., and Lunney, E. A., *J. Med. Chem.* **34**, 1258–1271 (1991)

[42] Thaisrivongs, S., Pals, D. T., Turner, S. R., and Kroll, L. T., *J. Med. Chem.* **31**, 1369–1376 (1988)

[43] Rahuel, J., Priestle, J., and Gruetter, M. G., *J. Struct. Biol.* **107**, 227–236 (1991)

[44] Dhanaraj, V., Dealwis, C. G., Frazao, C., Badasso, M., Sibanda, B. L., Tickle, I. J., Cooper, J. B., Driessen, H. P. C., Newman, M., Aguilar, C., Wood, S. P., Blundell, T. L., Hobart, P. M., Geoghegan, K. F., Ammirati, M. J., Danley, D. E., O'Connor, B. A., and Hoover, D. J., *Nature* **357**, 466–472 (1992)

[45] Lim, L. W., Roderick, A. S., Leimgruber, N. K., Gierse, J. K., Abdel-Meguid, S. S., *J. Mol. Biol.* **210**, 239–240 (1989)

[46] Sielecki, A. R., Hayakawa, K., Fujinaga, M., Murphy, M. E. P., Fraser, M., Muir, A. K., Carilli, C. T., Lewicki, J. A., Baxter, J. D., and James, M. N. G., *Science* **243**, 1346–1351 (1989)

[47] Frazao, C., Topham, C., Dhanaraj, V., Blundell, T. L., *Pure Appl. Chem.* **66**, 43–50 (1994)

[48] Blundell, T. L., Carney, D., Gardner, S., Hayes, F., Howlin, B., Hubbard, T., Overington, J., Singh, D. A., Sibanda, B. L., and Sutcliffe, M. J., *Eur. J. Biochem.* **172**, 513–520 (1988)

[49] Toh, H., Ono, M., Saigo, K., and Miyata, T., *Nature* **315**, 691–692 (1985)

[50] Miller, M., Jaskolski, M., Rao, J. K. M., Leis, J., and Wlodawer, A., *Nature* **337**, 576–579 (1989)

[51] Katz, R. A., and Skalka, A. M., *Annu. Rev. Biochem.* **63**, 133–173 (1994)

[52] Kohl, N. E., Emini, E. A., Schleif, W. A., Davis, L. I., Heimbach, J. C., Dixon, R. A., Scolnick, E. M., and Sigal, I. S., *Proc. Natl Acad. Sci., USA* **85**, 4686–4690 (1988)

[53] McQuade, T. K., Tomasselli, A. G., Liu, L., Karacostas, V., Moss, B., Sawyer, T. K., Heinrikson, R. L., and Tarpley, W. G., *Science* **247**, 454–456 (1990)

[54] Kaplan, A. H., Zack, J. A., Knigge, M., Paul, D. A., Kempf, D. J., Norbeck, D. W., and Swanstrom, R. J., *J. Virol.* **67**, 4050–4055 (1993)

[55] Humblet, C. C., Lunney, E. A., Buckheit Jr., R. W., Doggett, C., Wong, R. and Antonucci, T. K., *Antiviral Research* **21**, 73–84 (1993)

[56] Weber, I. T., Miller, M., Jaskolski, M., Leis, J., Skalka, A. M., and Wlodawer, A., *Science* **243**, 928–931 (1989)

[57] Navia, M. A., Fitzgerald, P. M. D., McKeever, B. M., Leu, C.-T., Heimbach, J. C., Herber, W. K., Sigal, I. S., Drake, P. L., and Springer, J. P., *Nature* **337**, 615–620 (1989)

[58] Wlodawer, A., Miller, M., Jaskolski, M., Sathyanarayana, B. K., Baldwin, E., Weber, I. T., Selk, L. M., Clawson, L., Schneider, J., and Kent, S. B. H., *Science* **245**, 616–621 (1989)

[59] Appelt, K., *Perspect. Drug Discovery Des.* **1**, 23–48 (1993)

[60] Erickson, J. W., *Perspect. Drug Discovery Des.* **1**, 109–128 (1993)

[61] Kuntz, I. D., Blaney, J. M., Oatley, S. J., Langridge, R., and Ferrin, T. E., *J. Mol. Biol.* **161**, 269–288 (1982)

[62] DesJarlais, R. L., Seibel, G. L., Kuntz, I. D., Furth, P. S., Alvarez, J. C., Ortiz de Montellano, P. R., DeCamp, D. L., Babe, L. M., Craik, C. S., *Proc. Natl Acad. Sci., USA* **87**, 6644–6648 (1990)

[63] Rutenber, E., Fauman, E. B., Keenan, R. J, Fong, S., Furth, P. S., Oritz deMontellano, P. R., Meng, E., Kuntz, I. D., DeCamp, D. L., Salto, R., Rose, J. R., Craik, C. S., and Stroud, R. M., *J. Biol. Chem.* **268**, 15343–15346 (1993)

[64] Appelt, K., 33rd Interscience Conference on Antimicrobial Agents & Chemotherapy, New Orleans, LA., 1993

[65] Kim, E. E., Baker, C. T., Dwyer, M. D., Murcko, M. A., Rao, B. G., Tung, R. D., and Navia, M. A., *J. Am. Chem. Soc.* **117**, 1181–1182 (1995)

[66] Ramnarayan, S. R., Pan, W., Gulnik, S. V., Burt, S., and Erickson, J. W., *Biorg. Med. Chem. Lett.* **4**, 1247–1252 (1994)

[67] Lam, P. Y. S., Jadhav, P. K., Eyermann, C. J., Hodge, C. N., Ru, Y., Bacheler, L. T., Meek, J. L., Otto, M. J., Rayner, M. M., Wong, Y. N., Chang, C.-H., Weber, P. C., Jackson, D. A., Sharpe, T. R., and Erickson-Viitanen, S. K., *Science* **263**, 380–383 (1994)

[68] DesJarlais, R. L., Sheridan, R. P., Seibel, G. L., Dixon, J. S., Kuntz, I. D., and Venkataraghavan, R., *J. Med. Chem.* **31**, 722–729 (1988)

[69] Chenera, B., DesJarlais, R. L., Finkelstein, J. A., Eggleston, D. S., Meek, T. D., Tomaszek, Jr., T. A., and Dreyer, G. B., *Biorg. Med. Chem. Lett.* **3**, 2717–2722 (1993)

[70] Tummino, P. J., Ferguson, D., Hupe, L., and Hupe, D., *Biochem. Biophys. Res. Commun.* **200**, 1658–1664 (1994)

[71] Thaisrivongs, S., Tomich, P. K., Watenpaugh, K. D., Chong, K.-T., Howe, W. J., Yang, C.-P., Strohbach, J. W., Turner, S. R., McGrath, J. P., Bohanon, M. J., Lynn, J, C., Mulichak, A. M., Spinelli, P. A., Hinshaw, R. R., Pagano, P. J., Moon, J. B., Ruwart, M. J., Wilkinson, K. F., Rush, B. D., Zipp, G. L., Dalga, R. J., Schwende, F. J., Howard, G. M., Padbury, G. E., Toth, L. N., Zhao, Z., Koeplinger, K. A., Kakuk, T. J., Cole, S. L., Zaya, R. M., Piper, R. C., and Jeffrey, P., *J. Med. Chem.* **37**, 3200–3204 (1994)

[72] Lunney, E. A., Hagen, S. E., Domagala, J. M., Humblet, C., Kosinski, J., Tait, B. D., Warmus, J. S., Wilson, M., Ferguson, D., Hupe, D., Tummino, P. J., Baldwin, E. T., Bhat, T. N., Liu, B., and Erickson, J. W., *J. Med. Chem.* **37**, 2664–2677 (1994)

[73] Goodsell, D. S., and Olson, A. J., *Proteins: Struct., Funct. Genet.* **8**, 195–202 (1990)

[74] (a) Goodford, P. J., *J. Med. Chem.* **28**, 849–857 (1985); (b) GRID Software Program, Molecular Discovery Ltd., West Way House, Elms Parade, Oxford, OX2 9LL, UK

[75] Vara Prasad, J. V. N., Para, K. S., Lunney, E. A., Ortwine, D. F., Dunbar, Jr., J. B., Ferguson, D., Tummino, P. J., Hupe, D., Tait, B. D., Domagala, J. M., Humblet, C., Bhat, T. N., Liu, B., Guerin, D. M. A., Baldwin, E. T., Erickson, J. W., and Sawyer, T. K., *J. Am. Chem. Soc.* **116**, 6989–6990 (1994)

[76] Ortwine, D., Dunbar, Jr., J. B., Lunney, E. A., Vara Prasad, J. V. N., Para, K. S., Bhat, T. N., Liu, B., Baldwin, E. T., and Erickson, J. W. *A Comparison of Docking Approaches.* 10th European Symposium on Structure-Activity Relationships: QSAR and Molecular Modeling, Barcelona, 4–9 September (1994)

[77] Ho, C. M. W., and Marshall, G. R., *J. Comput.-Aided Molec. Des.* **4**, 337–354 (1990)

[78] Vara Prasad, J. V. N., Para, K. S., Tummino, P. J., Ferguson, D., McQuade, T. J., Lunney, E. A., Rapundalo, S. T., Batley, B. L., Hingorani, G., Domagala, J. M., Grachek, S. J., Bhat, T. N., Liu, B., Baldwin, E. T., Erickson, J. W., and Sawyer, T. K., *J. Med. Chem.* **38**, 898–905 (1995)

[79] Tait, B. D. *HIV Protease: Non-peptide Leads from Compound Collections.* The First Winter Conference on Medicinal and Bioorganic Chemistry, Steamboat Springs, Jan. 29–Feb. 2 (1995)

[80] Olson, A. J., and Morris, G. M., *Perspect. Drug Discovery Des.* **1**, 329–344 (1993)

[81] Blaney, J. M., and Dixon, J. S., *Perspect. Drug Discovery Des.* **1**, 301–319 (1993)

[82] Erickson, J. W., Neidhardt, D. J., VanDrie, J., Kempf, D. J., Wang, X. C., Norbeck, D. W., Plattner, J. J., Rittenhouse, J. W., Turon, M., Wideburg, N., Kohlbrenner, W. E., Simmer, R., Helfrich, R., Paul, D. A., and Knigge, M., *Science* **249**, 527–533 (1990)

4 Zinc Endoproteases: A Structural Superfamily

N. Borkakoti

Abbreviations

ACE   Angiotensin Converting Enzyme
BoNt Botulinum Neurotoxin
BMP   Bone Morphogentic Protein
ECE   Endothelin Converting Enzyme
NEP   Neutral endopeptidase
TeNt Tetanus Neurotoxin

4.1 Introduction

On the basis of the mode of action, zinc-dependent proteases may be broadly divided into three subfamilies:

1. Aminopeptidases which are exopeptidases that catalyze the removal of unsubstituted amino acids from the N-terminus of proteins;
2. Endopeptidases that form the majority of zinc-dependent metalloproteins; and
3. Carboxypeptidases which remove the C-terminal amino acids from proteins.

These families may, in general, be identified according to the conserved amino acids implicated in the binding of the catalytic zinc atom [1–5]. The crystal structures of various zinc metalloproteins, as summarized in Fig. 1, have served to reinforce the primary sequence consensus in terms of common three-dimensional topology within each subclass. For example, the crystal structures of pancreatic metallocarboxypeptidases A1 [6–8], B [9, 10] and A2 [11] show a similar protein fold [12] and the catalytic site of the enzymes contain an essential zinc coordinated by the residues consistent with the signature pattern HxxE proposed on the basis of primary sequence. The recent explosion of structural data [13–26] on several zinc endoproteases with varying function and sequence provides a basis for a topological study of this subset of hydrolases. The observations and conclusions derived from the comparative analyses of the structural data currently available on zinc metalloendoproteases is presented in this review.

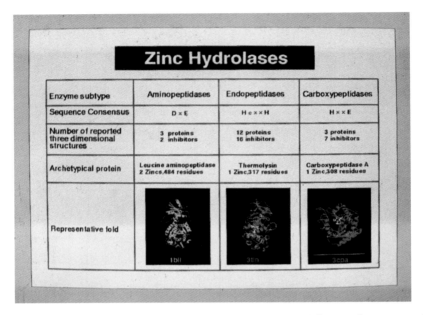

Figure 1. Classification and overview of the known structures of zinc metalloproteases. Amino acids (in one-letter code) of the signature sequence involved in binding the catalytic zinc are shown in black capital letters, variable residues are represented as x and residues implicated in catalysis but not involved in zinc binding are shown in red. The Brookhaven Databank [53] of the proteins used to represent archetypical folds are also shown.

4.2 Structural Classification of Zinc Endoprotease Families

Thermolysin, a thermostable calcium binding zinc hydrolase isolated from *Bacillus thermopro-teolyticus*, was the first zinc endopeptidase for which detailed structural information was available [27, 28] and has since been used as a prototype [29–31] for a variety of zinc proteases. The tertiary structure of thermolysin showed that the conserved zincin motif HexxH is part of an alpha-helix (Fig. 2) with the two histidine residues chelating the catalytic zinc, and the glutamic acid following the first histidine is ideally located to act as nucleophile during catalysis [32]. The identity and the number of intervening residues between the zinc ligands in the catalytic helix and the third zinc-binding group defines the hydrolases, the zincins [33], into two subfamilies: (1) Short spacer or metzincin [33] family and (2) Long spacer or gluzincin [5] family.

4.2.1 Short Spacer or Metzincins Family

Proteins which have the consensus HexxHxxgxxH, where the separation between the first and third (histidine) zinc ligands is nine residues. When the primary sequence in regions other than the zinc binding motif are considered, these proteases to be grouped into four [4, 5, 33] classes (Fig. 2). Since most of the zinc proteases are multimodular proteins, with indi-

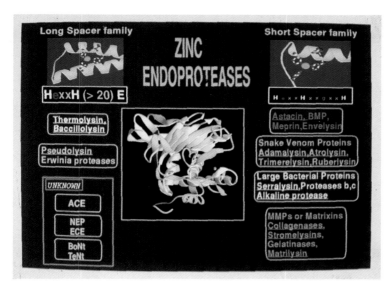

Figure 2. Structural classification of zinc endoproteases. The zinc binding ligands of the consensus sequence (one-letter amino acid code, black panels) are shown in white capitals, variable residues marked are as 'x', the essential catalytic residue is shown in red, and conserved residues in green. The same residue color code applies to the observed structure segments of the consensus motifs (blue panel) representative of the two categories. The location of the catalytic zinc atom is denoted by a cyan sphere. The central panel shows the superposition of the alpha carbon atoms of the topologically equivalent catalytic domains of proteins from the long consensus family (silver) and the short spacer family (orange). Enzymes for which three-dimensional structures are available are underlined, with selected proteins having related primary sequences enclosed in boxes. Protein families for which there are no representative structures are also identified. Zinc-dependent amino- and carboxy-peptidases possessing the HexxH motif are not included (see text).

vidual members of protein families having varying number of domains [19, 34, 35], their classification is based solely on the similarity of the mandatory catalytic domain of about 200 residues. The sequence homology in these regions is better than 35% between members within the same class.

Astacin [13], the crayfish endopeptidase, has an open-sandwich topology, with two alpha-helices folded against a twisted five-stranded beta-sheet (Fig. 3). The catalytic helix, located towards the C-terminus of the structure provides two histidines to ligate the zinc. The conserved glycine of the consensus HexxHxxgxxH provides a crucial turn in the protein which enables the third histidine to coordinate to the zinc (Figs. 2 and 3). These features are conserved (Fig. 4) in the catalytic domain of fibroblast collagenase [19], a member of the matrixin family, in spite of the low primary sequence homology between collagenase and astacin (Table 1). Although these proteins vary in detail, for instance, additional structural metal atoms observed in collagenase [17, 19, 20, 25] or presence of an extra loop past the catalytic helix in astacin, the preservation of the overall fold is striking. A pairwise comparison of the catalytic domain of of representative members [36] of each class of protein (Table 1) clearly shows the conservation of secondary structure in spite of very low sequence homology.

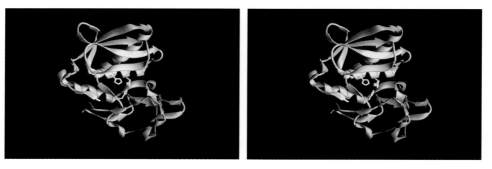

Figure 3. The observed fold for astacin. Stereo ribbon [54] drawing of the observed structure of astacin [13]. The catalytic zinc atom (cyan sphere) and the zinc binding histidine residues (blue) of the consensus sequence HexxHxxgxxH are identified. The structural elements defining the conserved zinc-endoprotease fold are shown in orange.

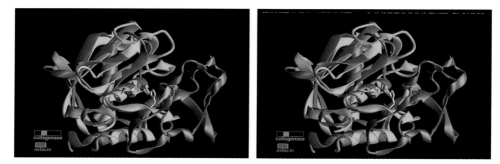

Figure 4. Comparison of astacin and the catalytic domain of collagenase. Stereo ribbon diagram [54] of the structure of astacin (green) and the catalytic domain of collagenase (orange) superimposed using a rigid body fit of the catalytic helices. The catalytic zinc atom (cyan) and the zinc ligands (blue) are marked. The overall similarity in the open beta sandwich region, N-terminal to the active site helix is apparent.

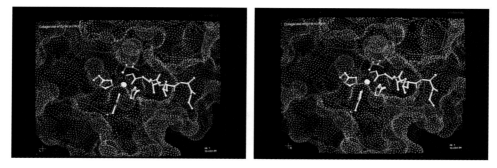

Figure 5. Connolly surface of truncated collagenase, complexed with **1**. Structure of the catalytic domain of fibroblast collagenase with a corresponding Connolly surface [55], shown in stereo. The active site cleft accommodating the catalytic zinc (white sphere) and the deep S1′ specificity pocket are clearly seen. The hydroxamate inhibitor (not included in the surface calculations) is shown in yellow. Protein residues involved in the catalysis are indicated.

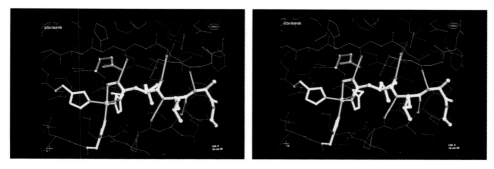

Figure 6. Collagenase–inhibitor **1** complex – a typical example of enzyme–inhibitor interactions for the short spacer family. A detailed stereo view of a synthetic hydroxamate containing inhibitor **1** (yellow ball and stick representation) [19] bound to the catalytic domain of collagenase (green). The catalytic zinc (cyan sphere) is coordinated by two oxygen atoms of the inhibitor and three histidine residues (white) of the protein. The glutamate involved in catalysis is shown in red. Putative hydrogen bonds are shown as orange cylinders. Atom coloring is red, blue and white for oxygen, nitrogen and carbon respectively.

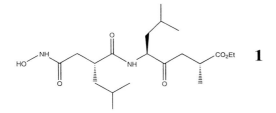

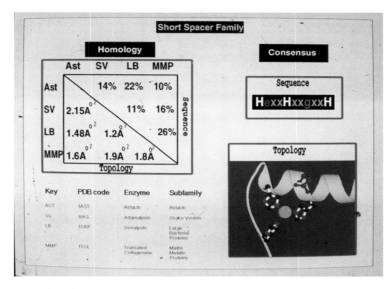

Table 1. Pairwise comparison of the topology and primary sequence of members of the short spacer family. The alpha carbon atoms defining the zinc protease fold (orange segment, Fig. 3) have been used in the topological superposition [56]. The distances refer to the root mean square deviations of this fold between pairs of structures. The corresponding pairwise primary sequence homology is also shown.

The overall structure of these metalloproteins is characterized by a deep active site cleft that divides the molecule into two domains, with the catalytically active zinc located at the bottom of the active site in the center of the molecule (Fig. 3). The structures of protein–inhibitor complexes [15, 17, 19–26] representing different families, indicate that the P1′ residue [37] is the principal recognition element for these enzymes (Fig. 5). Inhibitors bound to members of the matrixin family [17, 19–23, 25] adopt an extended conformation (Fig. 6), with the side chain at the P2′ subsite directed away from the enzyme on the opposite side of the inhibitor backbone to the P1′ specificity pocket. In addition to the zinc ligands, there are essential enzyme–inhibitor interactions between backbone atoms of the protein and complementary atoms of the inhibitor **1**.

4.2.2 Long Spacer or Gluzincins Family [5]

Gluzincins are proteins with the consensus HexxH(> 20)E, where the third zinc ligand, a glutamate, lies at least 20 residues C-terminal to the zincin motif (Table 2). The archetypical protein of this class, thermolysin consists of two domains, with the active site located between the N-terminal catalytic domain and the all alpha C-terminal domain (Fig. 7). With the limited data available, a comparison of topological matches of the observed structures with their cor-

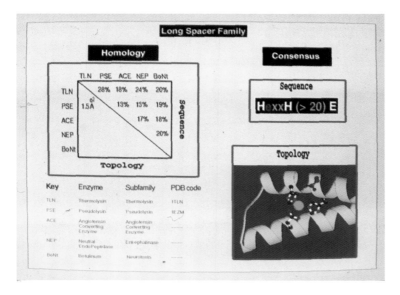

Table 2. Pairwise comparison of the topology and sequence of members of the long spacer family. Primary sequences for human Angiotensin Converting Enzyme [57], Botulinum Neurotoxin A [58], human Neutral Endopeptidase [59] have been used in the pairwise sequence comparison. Legend as in Table 1.

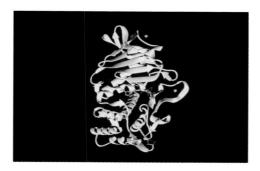

Figure 7. The structure of thermolysin. Ribbon representation of the structure of thermolysin (silver, Brookhaven Databank [53] code 3TLN) shown with a bound inhibitor (green). The catalytic zinc atom (cyan) and structural calcium atoms (magenta) are shown. The active site is located between the N-terminal zinc protease domain and the alpha helical C-terminal domain. Zinc binding residues are in blue and the residue assisting catalysis is shown in red.

responding sequence identities (Table 2) highlights the conservation of the three-dimensional fold. When the N-terminal domain is compared with the catalytic modules of the short spacer, metzincin, family (Fig. 8), it can be seen that the arrangement of secondary structural elements in the latter proteins is a conformational subset of the catalytic domain observed for the thermolysin family [19, 38]. The conservation of the common scaffold of the twisted beta-sheet, two helices and two of the three zinc-chelating ligands proteins with very distant in primary sequence (<15% sequence identity) lends support to the common ancestory of zinc endoproteases. Consequently zinc proteases of unknown structure (Fig. 2) are likely to have a similar open sandwich fold. This hypothesis has been successfully exploited for designing inhibitors of pharmaceutically relevant zinc-dependent proteins such as Angiotensin Converting Enzyme [39] and Enkephalinase [30, 39], details of which are discussed in the following section.

The structure of thermolysin–inhibitor complexes has shown the presence of a deep P1′ specificity pocket and the details of hydrogen bonding between enzyme and inhibitor [28]. Side-chain atoms of the enzyme are involved in interaction with the inhibitor amides and the inhibitor conformation is such that the subsites P1′ and P2′ both point into the enzyme core (Fig. 9).

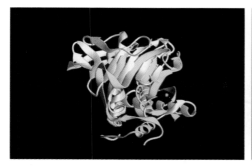

Figure 8. Superposition of the N-terminal domain of thermolysin (silver) and the catalytic domain of collagenase (orange). Stereo ribbon diagram illustrating the common topology between representative structures of the long spacer (silver) and short spacer (orange) families defining the zinc-endoprotease fold.

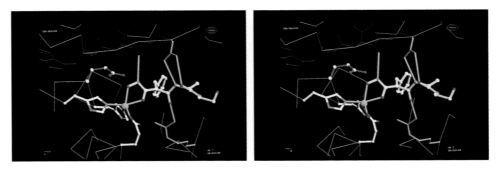

Figure 9. Thermolysin–inhibitor complex – an example of inhibitors bound to zinc endoproteases with the long spacer consensus. A stereo view of the active site site of thermolysin showing details of enzyme–inhibitor **2** interactions (Brookhaven Databank Code 5TLN). Enzyme side chains involved in inhibitor recognition are shown in magenta. The color code is as given for Fig. 6.

2

4.3 Overview of Inhibitor Design

An important feature of long spacer zinc-endoproteases like thermolysin, as revealed by comparisons of the free enzyme and that complexed with inhibitors [40, 41] is that conformational change is an essential component for catalytic activity. A similar conformational change is also probable [15, 18, 26] for the short spacer family, although a definite confirmation must await more structural information on equivalent free and inhibited proteins of this class.

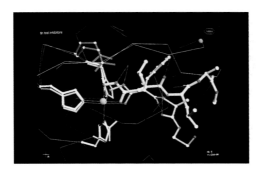

Figure 10. Comparison of the details of inhibitor binding in the two families: superposition of the catalytic centers of collagenase (Fig. 6) and thermolysin (Fig. 9). The catalytic zinc atom is shown as a cyan sphere, with the collagenase-inhibitor complex in yellow and the thermolysin complex shown in magenta. Thin cylinders denote putative hydrogen bonds. The green sphere shows the location of a structural calcium atom observed in collagenase. For clarity, only the alpha-carbon atoms of the proteins are shown. The differences in the relative oreintation of the subsites are evident.

In spite of the overall similarity of tertiary structure, a detailed analysis of the binding of inhibitors (Fig. 10), shows important differences in inhibitor-binding properties between the classes. In the short spacer family, the inhibitor is bound in an extended conformation while in the thermolysin family inhibitors adopt a twisted conformation. The contrasting requirements of the two classes is illustrated by the selectivity of the classical non-specific zinc-protease inhibitor phosphoramidon which is active in nanomolar concentrations against the thermolysin-like enzymes [28] but has little or no inhibitory activity against the enzymes of the short spacer family [42].

The difference in the orientation of the subsites for the two families can be seen in the potencies of specific cyclic inhibitors for each subgroup. In the long-spacer family, the contiguous orientation of the P1' and P2' pockets, as observed for the structures of thermolysin [28] and pseudolysin [41], has formed the basis for the design of novel inhibitors of Enkephalinase (NEP or neutral endopeptidase) [29] and Angiotensin Converting Enzyme (ACE) [30, 39], members of the gluzincin family for which the three-dimensional structures are not determined (Fig. 2). Computer modeling studies on the binding of inhibitors to the active center of thermolysin led to the design of 10- to 13-membered ring macrolactams, joining the P1' and P2' subsites. Incorporation of the mandatory zinc-chelating moiety to such cyclic lactams resulted in inhibitors active at nanomolar concentrations against both ACE and NEP [39] (Fig. 11). Alteration of the ring size and changes of the functionality in the P3' subsite proved effective in controlling selectivity.

In contrast, for inhibitors interacting with members of short consensus zinc endoprotease family, like the matrixins [17, 19–23, 25], the subsites P2' and P3' are suitably orientated for bridging. Molecules having the essential zinc-binding functionality and macrolactams span-

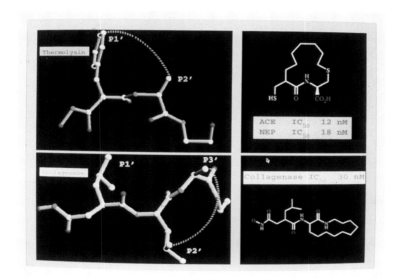

Figure 11. Cyclic inhibitors **3** and **4** of zinc-endoproteases. The differences in the conformation of the inhibitors leads to separate series of inhibitors for the two families. For details, see text.

ning subsites C-terminal to the P1′ specificity pocket, represent potent inhibitors for this subset of enzymes (Fig. 11). It is expected that, analogous to the gluzincin family, the potency and efficacy of this class of compounds against the metzincins, may be regulated by modifying the exocyclic components.

Selectivity between enzymes within family closely related in sequence has been achieved by exploiting the small changes in primary sequence which translate into large differences in the three-dimensional structure. The design of inhibitors for members of the Matrix Metallo Protease (MMP) family (Table 3), for example, is dictated by the sequence changes affecting the specificity pocket P1′. As illustrated in Fig. 5, matrixins are characterized by a deep P1′ pocket. In the observed structure of fibroblast collagenase complexed with a hydroxamate inhibitor [19] it can be seen that the walls of the specificity pocket are formed by the residues from the catalytic helix on one side (left, Fig. 12A(i)) and the side chains around the 'Met-turn' [33, 43] on the other. A buried arginine residue at the base of this subsite limits the size of the P1′ pocket in fibroblast collagenase (Fig. 12). In all other MMPs (Table 3, marked p1′) the pocket is larger, due to replacement of this arginine by shorter side chains. For example, the substitution of a leucine in stromelysin for the arginine in fibroblast collagenase results in stromelysin having a considerably deeper P1′ pocket relative to collagenase. Consequently, the subsite in stromelysin is able to accomodate inhibitors with longer side chains at P1′ more easily than fibroblast collagenase (Fig. 12B(i)). Thus, inhibitors with long alkyl chains at P1′ (Fig. 12B(ii)) are selective inhibitors of stromelysin, being active at nanomolar concentrations against stromelysin but having a low (micromolar) activity towards fibroblast collagenase [44, 45].

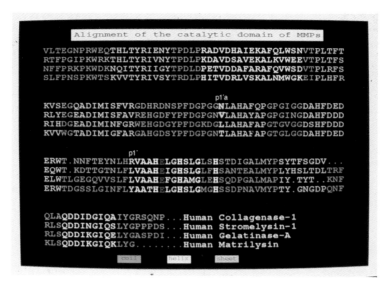

Table 3. Alignment of the sequences of the catalytic domain of selected human matrix metallo proteases (Matrixins). The sequences have been colored according to the secondary structure, with the zinc-chelating residues shown in white and the catalytic glutamic acid shown in red. Yellow residues identify amino acids discussed in the text.

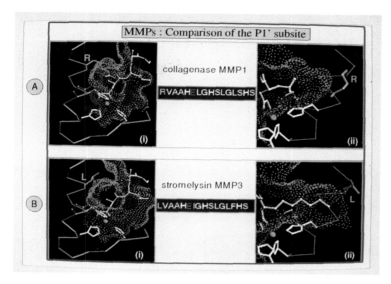

Figure 12. Comparison of the specificity pockets in matrixins. Two views of the specificity pocket in (A) fibroblast collagenase (MMP1) and (B) stromelysin (MMP3), illustrating the size of the P1′ subsite. The Connolly surface of the enzyme within 8 Å of the P1′ residue of the inhibitor (yellow) is shown together with the alpha carbon atoms of the catalytic helix (green). The residues are color-coded as given for Fig. 6 with the side chains discussed in the text shown in magenta. For clarity, only part of the inhibitor is shown. The data on the stromelysin complexes have been derived by homology model building (using Moloc [56]) on the corresponding data on collagenase. The relevant sequence segments of the catalytic helix of human fibroblast collagenase and human stromelysin are included.

Figure 13. A stereo view showing the details of enzyme inhibitor interactions between fibroblast collagenase (MMP1) and the synthetic inhibitor RO32-0554 (**5**). The color code is as given for Fig. 6, with additional enzyme side chains discussed in the text coloured magenta.

In an analogous manner, inhibitors bearing substituents at the carbon spacer between the zinc ligand and the alpha centre of the P1′ subsite have different enzyme–inhibitor interactions with separate members of the MMP family. The crystal structure of the synthetic inhibitor RO32-0554 (**5**) (Fig. 13) complexed with fibroblast collagenase indicates a favorable interaction between the imide oxygen of the naphthalimide group of the inhibitor and an

5

asparagine side chain of the enzyme (Table 3, marked p1'a). In stromelysin and gelatinase this asparagine is changed to a valine and leucine respectively, so that a similar interaction with the imide moiety of RO32-0554 (**5**) is not possible for these proteins. The changes in the enzyme–inhibitor interactions are reflected in the efficacy of inhibitor [44, 45] which is more effective against collagenase (IC_{50} 0.5 nM) than against stromelysin (IC_{50} 37 nM) or gelatinase (IC_{50} 5 nM).

4.4 Current Limitations

The presence of the sequence HexxH in the primary structure of a protein has been used to identify zinc-dependent proteases. Several proteins, such as tetanus [46] and botulinum toxins [47], leukotriene A4 hydrolase [48], were discovered to possess zinc-dependent peptidase activity after the identification of this consensus in their primary sequence. However, the X-ray structure of the enterotoxin C2 (SEC2) from bacteria [49], a superantigen with no known proteolytic or other hydrolytic activity illustrates the fallibilty of sequence-based classification. In SEC2, the histidines of the consensus signature HexxH provide two ligands to a zinc atom, but unlike the zinc endopeptidases where the motif HexxH is part of a helix, in SEC2 the sequence HexxH is a loop. The conformation and orientation of the glutamate essential for zinc-mediated catalysis is also significantly different [49]. Additionally, in contrast to the zincins, the third zinc ligand (aspartate) is towards the N-terminus of the first histidine. The presence of a fourth zinc ligand (aspartate) provided by a neighboring molecule in the crystal lattice suggests a non-catalytic role for the zinc ion in SEC2. The zinc-binding residues in SEC2 are not conserved in the family of staphylococcal enterotoxins with the second histidine of the consensus motif has been replaced by residues which have low affinity for zinc (lysine, arginine) [49]. This emphasizes the importance of analyzing sequences of homologous proteins in order to establish conserved sequences, since motifs common to a family of closely related proteins are more likely to be functional markers.

The structural classification of the zincins discussed above has been based on the three-dimensional data on zinc proteases currently available. These data (Fig. 2) are from zinc-dependent endopeptidases. It is therefore not mandatory for these enzymes to interact with the terminii of their protein substrates. However, for the zinc-dependent amino- and carboxypeptidases that possess the HexxH consensus [3–5], this additional requirement for substrate recognition may result in a three-dimensional structure which is considerably modified from the astacin scaffold observed for the endoproteases.

4.5 Future Prospects

The available three-dimensional structures of zinc endopeptidases with the consensus HexxH motif have shown remarkable topological equivalence of their secondary structures. Extension of the primary sequence consensus, to include the third zinc-binding ligand, divides these proteases into two major subfamilies, both in terms of primary sequence homology and details of secondary structure (Fig. 2). Screening the current database of primary sequences (SwissProt, February 1995) for the sequence HexxH shows that there are over 650 proteins with this signature. These enzymes have been classified as zinc endopeptidases, although the nature and location of the third zinc ligand are not known for all these proteins. Histidines, glutamic and aspartic acids and cysteines are the potential third ligand [50] for the

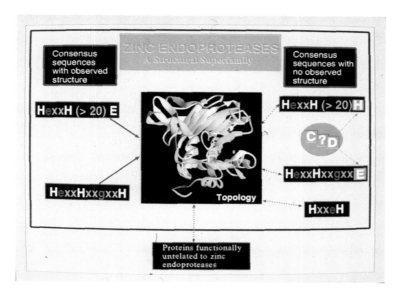

Figure 14. Zinc endoproteases: a structural superfamily. The central panel shows the superposition of the alpha carbon atoms of the topologically equivalent catalytic domains of proteins from the long consensus family (silver) and the short spacer family (orange). Various consensus sequences are given (see text) with the zinc-chelating residues shown in white. Full blue arrows indicate presence of motif in observed three-dimensional structures corresponding to the zinc endoprotease fold shown in the central panel. Dotted lines indicate putative structure–sequence relationships discussed in the text.

catalytic zinc. It is apparent that there are enzymes in the database which have sequence segments such as HexxHxxgxxE, HexxHxxgxxD. In addition, there are proteins with the sequence HexxH, which have histidine residues downstream of this motif, allowing these proteins to conform to an altered long consensus signature HexxH($>$20)H. However, predicting details of the secondary structure for proteases with 'hybrid' motifs is more challenging (Fig. 14).

Several additional points have to be considered:

(1) The proximity of the third ligand as the main determinant for the details of the fold. Details of the tertiary structure would then comply with the short spacer and long spacer delineation.
(2) The identity of the ligands to the catalytic metal has the principal influence on the secondary structure, implying that the net charge on the zinc atom (more positive in the short spacer family) would dictate the structural subset of the protein of interest.
(3) Unknown elements of primary sequence and zinc consensus determine the tertiary fold, indicating that the parallel structural and sequence categories classifing the zinc proteases is a result of the fortuitious selection of the proteases used for structural analysis.

The growing family of known divalent cation-dependent proteases such as insulinase [51] and dibasic convertase [52], with the variant consensus HxxeH, also present interesting questions as to the possibility of a 'mirrored' active site with or without conservation of the overall topology. Conversely, it is possible that entirely different proteins which have no zinc dependency and completely separate function may adopt the zinc endoprotease topology, simply because this fold provides a stable modular scaffold useful in the construction of multidomain proteins. Results of further structural studies are eagerly awaited.

References

[1]   Jongeneel, C. V., Bouvier, J., and Bairoch, A., *FEBS Lett.* **242**, 211–214 (1989)
[2]   Vallee, B. L., and Auld, D. S., *FEBS Lett.* **257**, 138–140 (1989)
[3]   Rawlings, N. D., and Barrett, A. J., *Biochem. J.* **275**, 389–391 (1991)
[4]   Jiang, W., and Bond, J. S., *FEBS Lett.* **312**, 110–114 (1992)
[5]   Hooper, N., *FEBS Lett.* **354**, 1–6 (1994)
[6]   Rees, D. C., Lewis, M., and Lipscomb, W. N., *J. Molec. Biol.* **168**, 367–387 (1983)
[7]   Kim, H., and Lipscomb, W. N., *Biochemistry* **29**, 5546–5555 (1990)
[8]   Kim, H., and Lipscomb, W. N., *Biochemistry* **30**, 8171–8180 (1991)
[9]   Coll, M., Guasch, A., Aviles, F. X., and Huber, R., *EMBO J.* **10**, 1–9 (1991)
[10] Aviles, F. X., Vendrell, J., Guasch, A., Coll, M., and Huber, R., *Eur. J. Biochem.* **211**, 381–389 (1993)
[11] Faming, Z., Kobe, B., Stewart, C., Rutter, W., and Goldsmith, E. J., *J. Biol. Chem.* **266**, 24606–24612 (1991)
[12] Murzin, A. G., Brenner, S. E., Hubbard, T., and Chothia, C., *J. Molec. Biol.* **247**, 536–540 (1995)
[13] Bode, W., Gomis-Ruth, F. X., Huber, R., Zwilling, R., and Stocker, W., *Nature* **358**, 164–167 (1992)
[14] Gomis-Ruth, F. X., Kress, L. F., and Bode, W., *EMBO J.* **12**(11), 4151–4157 (1993)
[15] Baumann, U., Wu, S., Flaherty, K. M., and McKay, D. B., *EMBO J.* **12**(9) 3357–3364 (1993)
[16] Hamada, K., Hiramatsu, H., Katsuya, Y., Hata, Y., Matsuura, Y., and Katsube, Y., *Acta Crystallogr. Sect. A* **49**, 153–157 (1993)
[17] Bode, W., Reinemer, P., Huber, R., Kleine, T., Schcnier, S., and Tschesche, H., *EMBO J.* **13**, 1263–1269 (1993)

[18] Baumann, U., *J. Molec. Biol.* **242**, 244–252 (1994)

[19] Borkakoti, N., Winkler, F. W., Williams, D. H., D'Arcy, A., Broadhurst, M. J., Brown, P. A., Johnson, W. H., and Murray, E. J., *Nature Struct. Biol.* **1**, 106–110 (1994)

[20] Lovejoy, B., Cleasby, A., Hassell, A. M., Longley, K., Luther, M. A., Weigl, D., McGeehan, G., McElroy, A. B., Drewry, D., Lambert, M. H., and Jordon, S. R., *Science* **263**, 375–377 (1994)

[21] Stams, T., Spurlino, J. C., Smith, D. L., Wahl, R. C., Ho, T. F., Qoronfleh, M. W., Banks, T. M., and Rubin, B., *Nature Struct. Biol.* **1**, 119–123 (1994)

[22] Gooley, P. R., O'Connell, J. F., Marcy, A. L., Cuca, G. C., Salowe, S. P., Bush, B. L., Hermes, J. D., Esser, C. K., and Hagmann, W. K., *Nature Struct. Biol.* **1**, 111–118 (1994)

[23] Browner, M. F., Smith, W. W., and Castelhano, A., *Biochemistry* **34**, 6602–6610 (1995)

[24] Zhang, D., Botos, I., Gomis-Ruth, F. X., Doll, R., Blood, C., Njoroge, F. G., Fox, J. W., Bode, W., and Meyer, F., *Proc. Natl Acad. Sci. USA* **91**, 8447–8451 (1994)

[25] Li, J., Brick, P., O'Hare, M. C., Skarzynski, T., Lloyd, L. F., Curry, V. A., Clark, I. M., Bigg, H. F., Hazelman, B. L., Cawston, T. E., and Blow, D., *Structure* **3**(6), 541–549 (1995)

[26] Baumann, U., Bauer, M., Letoffe, S., Delepelaire, P., and Wandersman, C., *J. Molec. Biol.* **248**, 653–661 (1995)

[27] Matthews, B. W., Weaver, L. H., and Kester, W. R., *J. Biol. Chem.* **249**(24), 8030–8044 (1974)

[28] Holland, D. R., Barclay, P. L., Danilewicz, J. C., Matthews, B. W., and James, K., *Biochemistry* **33**, 51–56 (1994)

[29] Johnson, W. H., Roberts, N. A., and Borkakoti, N., *J. Enzyme Inhibition* **2**, 1–22 (1987)

[30] Macpherson, L. J., Bayburt, E. K., Capparelli, M. P., Bohacek, R. S., Clarke, F. H., Ghai, R. D., Sakane, Y., Berry, C. J., Peppard, J. V., and Trapani, A. J., *J. Med. Chem.* **36**, 3821–3828 (1993)

[31] Beaumont, A., Le Moul, H., Boileau, G., Crine, P., and Roques, B. P., *J. Biol. Chem.* **266**(1), 214–220 (1991)

[32] Kester, W. R., and Matthews, B. W., *Biochemistry* **16**, 2506–2516 (1977)

[33] Bode, W., Gomis-Ruth, F. X., and Stocker, W., *FEBS Lett.* **331**, 134–140 (1993)

[34] Wolfsberg, T. G., Bazan, J. F., Blobel, C. P., Myles, D. G., Primakoff, P., and White, J. M., *Proc. Natl Acad. Sci. USA* **90**, 10783–10787 (1993)

[35] Stocker, W., Gomis-Ruth, F. X., Bode, W., and Zwilling, R., *Eur. J. Biochem.* **214**, 215–231 (1993)

[36] Stocker, W., Grams, F., Baumann, U., Gomis-Ruth, F. X., McKay, D. B., and Bode, W., *Protein Sci.* **4**, 823–840 (1995)

[37] Schechter, I., and Berger, A., *Biochem. Biophys. Res. Commun.* **27**, 157–162 (1967)

[38] Blundell, T. L., *Nature Struct. Biol.* **1**, 73–75 (1994)

[39] Stanton, J., Sperbeck, D. M., Trapani, A. J., Cote, D., Sakane, Y., Berry, C. J., and Ghai, R. D., *J. Med. Chem.* **36**, 3829–3833 (1993)

[40] Holland, D. R., Tonrud, D. E., Pley, H. W., Flaherty, K. M., Stark, W., Jansonius, J. N., McKay, D. B., and Matthews, B. W., *Biochemistry* **32**, 11310–11316 (1992)

[41] Thayer, M. M., Flaherty, K. M., and McKay, D. B., *J. Biol. Chem.* **266**, 2864–2871 (1991)

[42] Ye, Q. Z., Johnson, L. L., Hupe, D. J., and Baragi, V., *Biochemistry* **31**, 11231–11235 (1992)

[43] Grams, F., Reinemer, P., Powers, J. C., Kliene, T., Pieper, M., Tschesche, H., Huber, R., and Bode, W., *Eur. J. Biochem.* **228**, 830–841 (1995)

[44] Brown, P. A., Borkakoti, N., Bottomley, K. M., Broadhurst, M. J., D'Arcy, A., Hallam, T. J., Johnson, W. H., Lawton, G., Lewis, E. J., Murray, E. J., Williams, D. H., and Winkler, F. W., *ACS Abstracts* **207**, 1–2 (1994)

[45] Brown, P. A., Borkakoti, N., Bottomley, K. M., Broadhurst, M. J., D'Arcy, A., Hallam, T. J., Johnson, W. H., Lawton, G., Lewis, E. J., Murray, E. J., Williams, D. H., and Winkler, F. W., *J. Cell. Biochem. Suppl.* **18D**, 120 (1994)

[46] Schiavo, G., Poulin, B., Rossetto, O., Benfenati, F., Tauc, L., and Montecucco, C., *EMBO J.* **11**, 3577–3583 (1992)

[47] Montecucco, M., and Schiavo, G., *J. Biol. Chem.* **267**, 23479–23483 (1992)

[48] Medina, J. F., Wetterholm, A., Radmark, O., Shapiro, R., Haeggstrom, J. Z., Vallee, B. L., and Samuelsson, B., *Proc. Natl Acad. Sci. USA* **88**, 7620–7624 (1991)

[49] Papageorgiou, A. C., Acharya, K. R., Shapiro, R., Passalacqua, E. F., Brehm, R. D., and Tranter, H. S., *Structure* **3**, 769–779 (1995)

[50] Vallee, B. L., and Auld, D. S., *Proc. Natl Acad. Sci. USA* **87**, 220–224 (1990)

[51] Becker, A. B., and Roth, R. A., *Proc. Natl Acad. Sci. USA* **89**, 3835–3839 (1992)

[52] Pierotti, A. R., Prat, A., Chesneau, V., Gaudoux, F., Leseney, A. M., Foulon, T., and Cohen, P., *Proc. Natl Acad. Sci. USA* **91**, 6078–6082 (1994)

[53] Bernstein, F. C., Koetzle, T. F., Williams, G. J., Meyer, E. E., Brice, M. D., Rodgers, J. R., Kennard, O., Shimanouchi, T., and Tasumi, M., *J. Molec. Biol.* **112**, 535–542 (1977)

[54] Carson, M., *J. Molec. Graphics* **5**, 103–106 (1987)
[55] Connolly, M. J., *J. Appl. Crystallogr.* **16**, 548–558 (1983)
[56] Gerber, P. R., *Biopolymers* **32**, 1003–1017 (1992)
[57] Soubrier, F., Alhenc-Gelas, F., Hubert, C., Allegrini, J., John, M., Tregear, G., and Corvol, P., *Proc. Natl Acad. Sci. USA* **85**, 9386–9390 (1988)
[58] Thompson, D. E., Brehm, J. K., Oultram, J. D., Swinfield, T. J., Shone, C. C., Atkinson, T., Melling, J., and Minton, N. P., *Eur. J. Biochem.* **189**, 73–81 (1990)
[59] Malfroy, B., Kuang, W.-J., Seeburg, P. H., Mason, A. J., and Schofield, P. R., *FEBS Lett.* **229**, 206–210 (1988)

5 Structure-Based Design of Potent Beta-Lactamase Inhibitors

K. Gubernator, I. Heinze-Krauss, P. Angehrn, R.L. Charnas, C. Hubschwerlen, C. Oefner, M.G.P. Page and F. K. Winkler

5.1 Introduction

The treatment of infectious diseases still constitutes one of the main fields of modern pharmacology. Infections caused by Gram-negative bacteria contribute about half of the cases of life-threatening nosocomial diseases [1]. In recent years the frequency of resistance to modern beta-lactam antibiotics, particularly the third-generation cephalosporins, has increased exponentially [2]. In many cases, the resistance can directly be attributed to a high level of expression of class C beta-lactamases [3]. These enzymes are mostly chromosomally encoded but recently class C enzymes have appeared on plasmids [4]. Available types of beta-lactamase inhibitors such a clavulanate **1**, sulbactam **2** and tazobactam **3** (Fig. 1) are ineffective against these beta-lactamases and are therefore not suitable for use against these emergent organisms [3, 4]. Clavulanate and the penam sulfones undergo chemical rearrangements that are triggered by the attack of the enzyme on the beta-lactam ring and which eventually result in acrylic esters that are intrinsically more stable to hydrolysis [5]. The inefficiency of these compounds as inhibitors of the class C enzymes stems from the fact that hydrolysis of the initial intermediate (deacylation) can occur more rapidly than the chemical rearrangement to the acrylic ester in the class C active site [5]. We set out to use our structural and kinetic information about class C beta-lactamases as a basis for the design of mechanism-based inhibitors of this refractory class of enzymes that would specifically block the deacylation reaction.

5.2 Structure of *Citrobacter freundii* Class C Beta-Lactamase

The starting point for inhibitor design was the solution of the crystal structure of *Citrobacter freundii* 1203 beta-lactamase at 2.0 Å [6]. An overview of the structure is shown in Fig. 2. The active site of this enzyme is located in the center at the interface of two domains, left and right, both of which contribute catalytic residues.

In order to elucidate the individual role of the catalytic residues, we followed the example of Herzberg and Moult [7] and superimposed the native class *C. freundii* enzyme structure onto trypsin (PDB code 2PTC) using the active-site serine and the putative oxy-anion hole adjacent to it as reference points. A striking result of this superposition is that the phenolic oxygen of Tyr150 comes very close to the Nε of the essential histidine in trypsin. This suggested that Tyr150, as its anion, could act as a general base during catalysis of beta-lactam hydrolysis in a way similar to that of His57 in trypsin [6]. This hypothesis is still under examination by mutational data and kinetic investigations [8, 10].

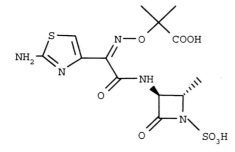

Clavulanate **1**   Sulbactam **2**

Tazobactam **3**   Penicillin G   **4**

Aztreonam **5**

Figure 1. Chemical structures substrates and inhibitors **1**–**5**.

5.3 Model of the Mechanism of Action: Cleavage of Penicillin G

In order to elucidate the mechanism of action of this enzyme in detail, penicillin G was selected as a model substrate for natural beta-lactams, such as penicillins and cephalosporins.

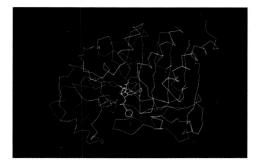

Figure 2. C-α display of the structure of *C. freundii* beta-lactamase. The active site residue side chains are drawn in blue. A model of Penicillin G **4** is displayed in yellow.

Models were built (Figs. 3 and 4) of the Michaelis complex (3b), the hemiacetal transition state (3c), the acyl enzyme (3d) and the deacylation step. All of these models could nicely be accommodated in the crystal structure without changing any protein coordinates except for those of the catalytic serine.

In the case of penicillin G, the outward rotation about the C3-C4 bond can occur without conflict with protein side chains. It relaxes the strained conformation along this bond previously enforced by the four-membered ring. Moreover, this rotation improves hydrophobic contact of the dimethyl and the sulfur part of the thiazolidine ring of penicillin with two leucine side chains, thus offering an additional driving force for the outward rotation. Experimental evidence is provided by the observed rotation of 35.2° in this direction about the C3-C4 bond that occurs in the acyl–enzyme complex formed between benzyl penicillin and a mutant class A beta-lactamase (RTEM-1) blocked in deacylation [9].

5.4 Structure of the Complex with Aztreonam

The structure of the acyl–enzyme complex formed with aztreonam **5** was solved at 2.5 Å resolution [6]. Comparison of the aztreonam moiety observed in the complex with the intact beta-lactam molecule indicated that rotation about the C3-C4 and N1-C4 bonds of respectively 70 and 45° had occurred. The rotation about C3-C4 relaxes the eclipsed conformation of the intact beta-lactam to the energetically more favorable gauche form (Fig. 5). Rotation in the opposite, outward direction, to adopt the alternative gauche conformation, is less favorable because it would bring the C4 methyl substituent into steric conflict with the side chain of Tyr150 (see Figs. 5 and 6) and Leu119. The rearrangement in the aztreonam complex leaves N1 and the sulfonate group in a position where they block one face of the ester bond formed with Ser64 (Fig. 5). They also displace a water molecule that, in the native enzyme structure, is in a position where it interacts with an extensive hydrogen-bonded network of residues that is an essential part of the catalytic mechanism of class C beta-lactamases. The residues involved in this network, Lys73, Tyr150, Asp152, and Lys315, all contribute somewhat to activation of water for hydrolysis of the acyl–enzyme complex [6, 10]. It was postulated that this water, positioned near Tyr150, is involved in deacylation [6]. It would be well-located for attack on the ester in the aztreonam complex, were it not displaced, and its access not blocked, by the N1-sulfonate.

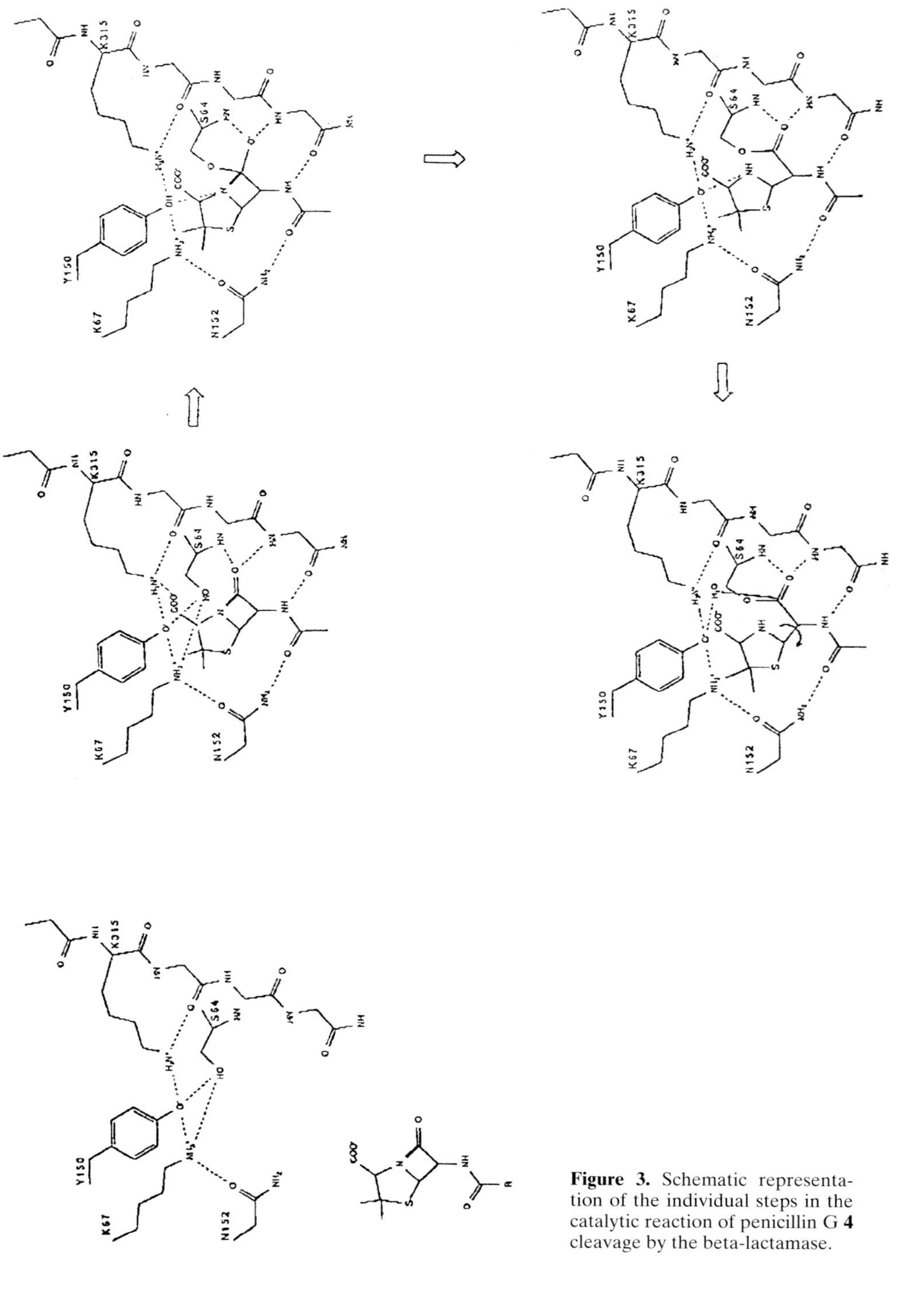

Figure 3. Schematic representation of the individual steps in the catalytic reaction of penicillin G **4** cleavage by the beta-lactamase.

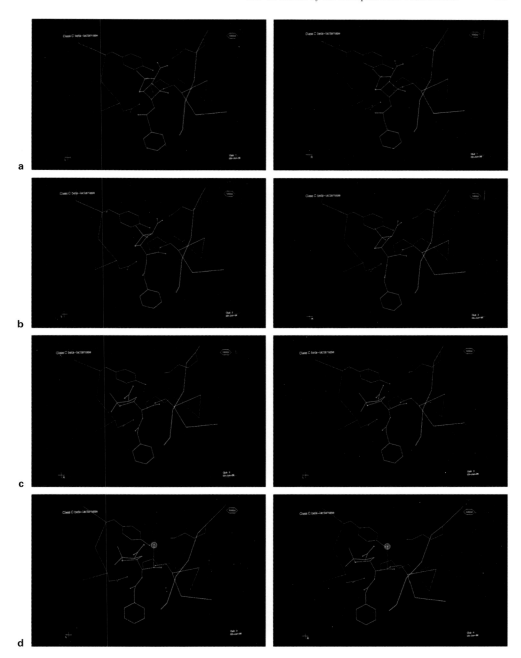

Figure 4. 3-Dimensional models of the indiviual steps in the catalytic reaction of penicillin G **4** cleavage by the beta-lactamase. The outward rotation occurs between the second and third image, offering unhindered access to the water molecule shown on the last image.

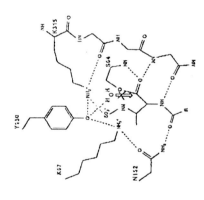

Figure 5. Schematic representation of the aztreonam **5** processing. Here, the reaction stops at the acylated stage and the deacylytion is hindered. The structure of the acylated state has been experimentally determined as shown in Fig. 6c.

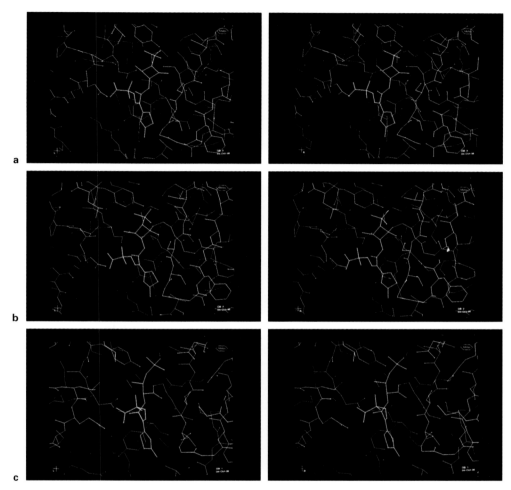

Figure 6. 3-Dimensional models of the aztreonam **5** processing. Here, the reaction stops at the acylated stage and the deacylation is hindered. The structure of the acylated state has been experimentally determined as shown in (c).

This experimental result offers a convincing rationalization for the slow deacylation of aztreonam in class C beta-lactamases.

5.5 Design of Inhibitors

Molecular modeling indicated that in class C enzymes, the outward rotation around the C3-C4 bond would open the path for the water molecule to attack the ester. Such compounds are rapidly hydrolyzed by the class C enzymes [11]. Therefore, preventing this rotation in a

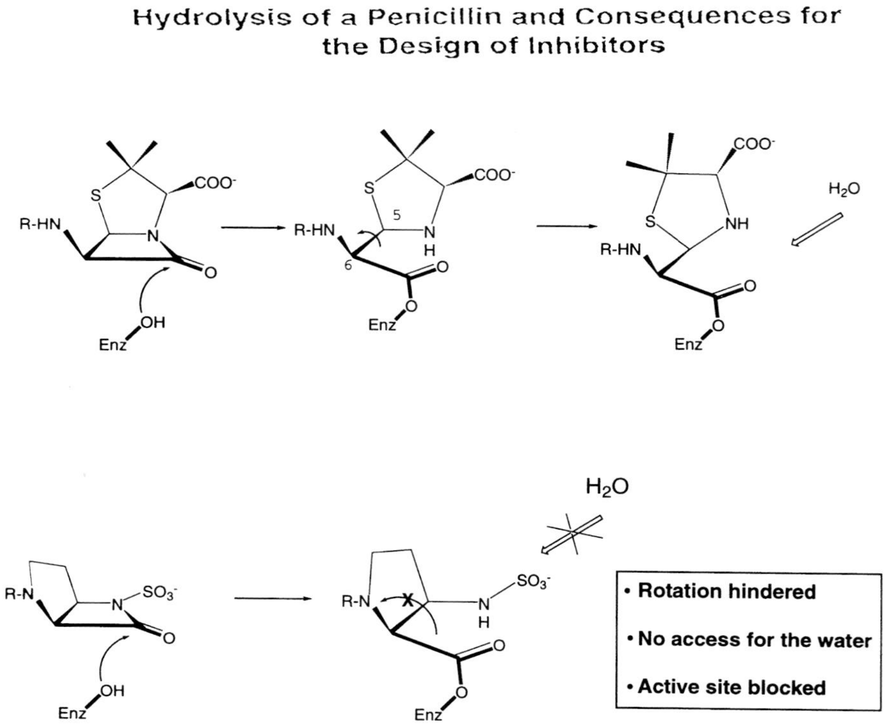

Hydrolysis of a Penicillin and Consequences for the Design of Inhibitors

Figure 7. Schematic representation of the design principles. The rotation around C3-C4 is hindered, access of water is blocked, the catalytic serine remains acylated, and the enzyme is inactivated.

suitable molecule should block the access of the water molecule to the ester bond and greatly stabilize the acyl–enzyme complex. This principle of restricted rotation blocking access for water and thus slowing down deacylation is shown in Fig. 7. We investigated this possibility by synthesizing monobactams that had N3 linked to C4 by a two-carbon atom bridge [12]. The inhibitors, which we term bridged monobactams, exhibited high affinity for a wide variety of class C beta-lactamases (Table 1). They inhibit the enzyme by rapidly acylating the catalytic serine and then remaining in this state without any significant deacylation, as predicted from the structural models.

5.6 Kinetics of the Inhibition Reaction

The reaction with class C beta-lactamases was typically biphasic, reflecting two kinetically distinct conformations of the protein [10, 13].

Table 1. Kinetic parameters describing the activity of selected bridged monobactams **6–9** in comparison with aztreonam (**5**).

| | Class C β-Lactamase | | | | Class A β-Lactamase | | | |
| | *Citrobacter freundii* 1982 | | | | RTEM-3 | | | |
Compound	IC_{50} μM	k_{acyl} s^{-1}	K_2 μM	k_{deacyl} $10^{-3} min^{-1}$	IC_{50} μM	k_{acyl} s^{-1}	K_S μM	k_{deacyl} $10^{-3} min^{-1}$
6: Ro 44-4454	0.11	0.23	245	4.2	> 10 mM	Not determined		
7: Ro 46-2800	0.02	0.59	11.6	0.08	5000	0.15	1200	190
8: Ro 46-2816	3.6	0.5	29.8	0.11	50	~0.1	~6940	180
9: Ro 46-7649	0.001	1.1	0.12	0.001	2.7	0.22	57	43
5: Aztreonam	0.053	0.74	105	8.5	3000	~1.5	3500	1100

$$E + I \iff E * I \xrightarrow{\;k_{acyl}\;} E - I \xrightarrow{\;k_{deacyl}\;} E + I^{\#}$$

with K_S above the first equilibrium.

The steady-state level of occupancy of the active site, which is dependent on the ratio of deacylation rate to the acylation rate (k_{deacyl}/k_{acyl}) [14], was greater than 0.999. The low deacylation rate was reflected in a very low net hydrolysis rate (Fig. 1), with $k_{cat}/K_M \le$ 1 $M^{-1} s^{-1}$.

The class A enzymes generally exhibit low affinity for the bridged monobactams, although some side chains did confer significant affinity (Table 1). At high concentration, acylation of

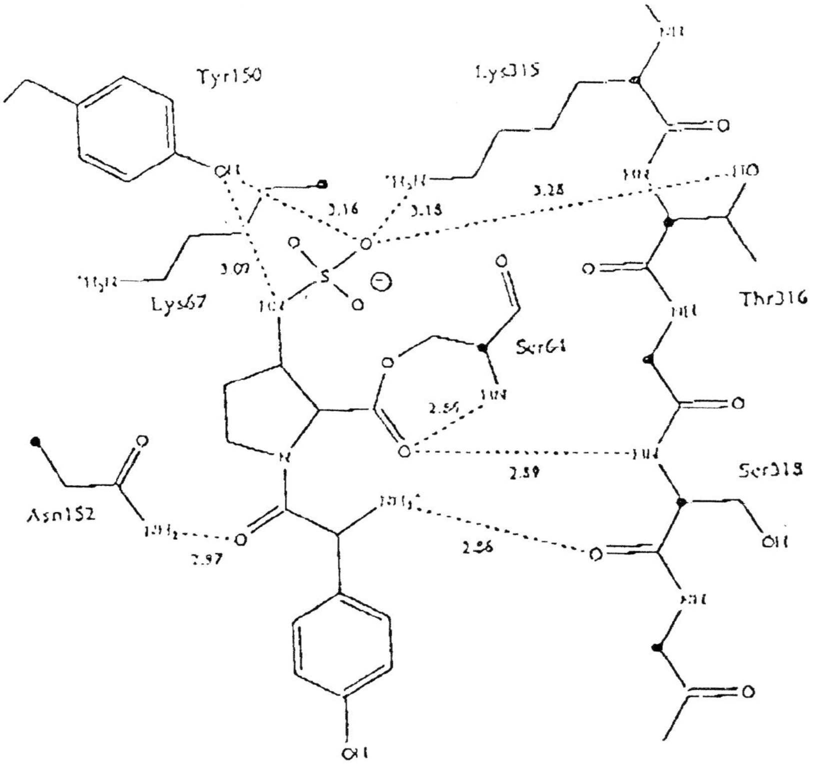

Figure 8a. Structure of the complex with the bridged monobactam Ro-48-2816 (**8**).

class A beta-lactamases was also rapid but the rate of breakdown of the acyl enzyme was also considerably higher. Thus, lower steady-state levels of occupancy of the active site and more rapid hydrolysis were observed, with $K_{cat}/K_M \leq 1$ M^{-1} s^{-1}.

5.7 Hydrolysis by Class A Beta-Lactamases

Inspection of class A beta-lactamase crystal structures indicated that a water molecule in an equivalent position would be much less activated because there is a serine residue replacing the tyrosine of class C enzymes [7, 11]. Attack in a class A enzyme occurs from the opposite side of the ester, with activation of the water molecule occurring through an extension of the hydrogen bond network that is not present in class C enzymes. Hence, rotation about C3-C4 is less critical to the mechanism of deacylation in these enzymes and restricting the rotation should have less impact on the stability of the acyl–enzyme complex.

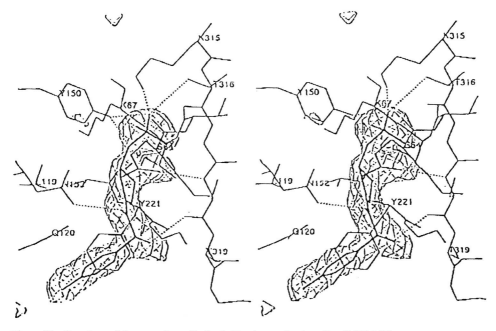

Figure 8b. Structure of the complex with the bridged monobactam Ro-48-2816 (**8**).

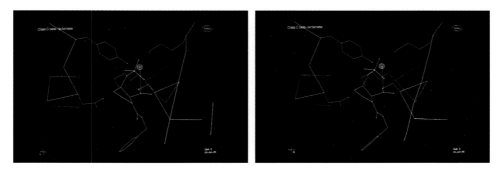

Figure 8c. Structure of the complex with the bridged monobactam Ro-48-2816 (**8**).

5.8 X-ray Structure of the Complex with a Bridged Monobactam 8

The stability of the acyl–enzyme complexes formed with *C. freundii* class C beta-lactamase allowed solution of their structures by X-ray crystallography (Fig. 8). No significant changes in protein structure occurred, except that the side chain of Asp123 moved to accommodate the inhibitor bound to a neighboring protein molecule. The conformation adopted by the inhibitor **8** shows that the beta-lactam ring has opened to form an ester which the catalytic serine residue. The *N*-sulfonate moiety forms a salt bridge with Lys315. N1 blocks the trajectory that an incoming water molecule would have to use for attack on the ester while the sulfonate displaces the water from the position where it can be activated. The carbonyl oxygen of the ester occupies the 'oxyanion hole', forming good hydrogen bonds to the main chain NH groups of Ser70 and Ser318. In this position, the reactivity of the ester towards attack of a water molecule should be comparable to that of a rapidly hydrolyzed substrate. Therefore, the stability of the acyl enzymes formed with the bridged monobactams is attributable solely to the denial of access to the water molecule. It does not involve deactivation as a result of either chemical rearrangements or moving the carbonyl oxygen out of the oxyanion hole.

A rotation about the amide bond of the acyl side chain by approximately 35°, compared to the aztreonam complex, is observed in the inhibitor complexes. This different vector is imposed by the pyrrolidine ring. It has the consequence of placing the side chain deeper in the active site. In turn, this restricts the options for side chain variation to compounds that have small substituents in the alpha-position and that are relatively flexible. Rigid side chains such as that of aztreonam are not compatible with these requirements and bridged monobactams with such side chains have very low affinities for the enzyme.

The side chain of the inhibitor lies parallel to the strand of the beta-sheet that forms one edge of the catalytic center. A hydrogen bond is possible between the carbonyl oxygen of the side chain and the amide nitrogen of Asn152, but its stereochemistry is far from ideal (Fig. 8). Nevertheless, this interaction appears to be important, as is illustrated by the affinity change in mutants of the *E. coli* class C beta-lactamase where Asn152 is replaced by either aspartate of leucine [10]. In these mutants, which have structure identical to the wild-type enzyme [10], the k_{acyl} (Table 1) for Ro-46-2800 (**7**) was increased from 13 µM to 1.26 mM. This corresponds to a change in binding energy ($\Delta\Delta G = RT \ln [K_{mut}/K_{wt}]$ of 11.9 kJ/mol. Loss of a direct hydrogen bond between amide nitrogen and amide carbonyl could result in a binding energy difference of between 12 and 38 kJ/mol [15, 16].

5.9 Structure–Activity Relationship among Bridged Monobactams

There are two other interactions that appear to influence side-chain specificity and affinity. First, there is the possibility to form a hydrogen bond between a donor, such as the alpha-amino group of Ro-46-2816 (**8**) (Fig. 8), and the backbone carbonyl group of Ser318 in the beta-strand. This interaction appears to make a considerably smaller contribution to the overall binding energy than that involving Asn152. K_M changes by less than a factor of 3 ($RT \ln [K_{c-x}/K_{c-h}]$ between -2.1 and 0 kJ/mol) among compounds with various substituents in the α-position [11], except when very bulky groups were introduced. This may be because the side chains of Tyr221, Val211 and Thr319 provide a rather apolar protein environment at this position that counteracts the favorable hydrogen bonding interaction. The second possibility is to form hydrophobic contacts with the aromatic ring of Tyr221, which lies at the bottom of the side-chain binding pocket. The influence of this interaction was clear in a series of compounds [12] having simple aromatic and aliphatic side chains (for example, Ro 46-2800

Table 2. Enzyme inhibitory constants (IC_{50}) and antibacterial minimum inhibitory concentration (MIC) results for synergy with ceftriaxone of a series of bridged monobactams.

R	IC_{50} (nM)			MIC ($\mu g\,mL^{-1}$) ceftriaxone:inhibitor 1:4		
	C. freundii 1982 (Class C)	*P. aeruginosa* 18 SH (Class C)	*E. coli* TEM-3 (Class A)	*C. freundii* 1982	*P. aeruginosa* 18 SH	*E. coli* TEM-3
D-(4-HOC_6H_4)CH(NH_2)	500	90	>100000	2	8	8
C_6H_5NH	220	420	NA	2	8	4
4-HOC_6H_4NH	109	155	>100000	1	8	8
3-(H_2NCO)C_6H_4NH	10	24	4000	1	16	16
4-(H_2NCO)C_6H_4NH	54	217	>100000	0.5	4	16
4-H_2NCONH)C_6H_4NH	12	48	100000	0.25	4	16
(lactam structure)	364	2246	>100000	1	4	16
(piperidine structure)	1350	4960	>100000	2	8	16
$C_6H_5CH_2O$	100	200	>100000	8	128	16
^tBuO	225	1500	>100000	8	8	16
(vinyl imidazole structure, E)	85	200	>100000	0.5	8	8
(H_2NCO pyridinium, N^+–CH_2 structure)	122	612	>100000	1	8	16
(thiazole S–CH_2, N–N structure)	283	853	>100000	1	128	8
(structure)	50	110	>100000	0.5	8	16
(structure)	23	27	>100000	0.5	4	8
(structure)	6	55	>100000	0.25	16	16
4-$HOC_6H_4NHCH_2$	3	25	79200	8	128	8
Reference ceftriaxone alone	—	—	—	128	128	16
Reference tazobactam	900	800	15	8	>32	0.25

(**7**) and Ro 46-2816 (**8**)). The binding energy was directly proportional to side-chain hydrophobicity, increasing by approximately 1.4 kJ/mol for a 1 kJ/mol change in hydrophobicity parameter π (calculated from literature values [17]).

Optimization of these interactions described above through modification of the acyl side chain increased the affinity over 10^4-fold, from a K_M value >10 mM (when none of the interactions are satisfied) to a K_M value of 120 nM (for Ro-46-7649; **9**). The stability of the acyl–enzyme complex also increased (Table 1), possibly because tighter binding of the inhibitor decreases the flexibility of the acyl–enzyme complex and thus the rate of occasional access of water to the ester.

A large number of such compounds have been synthesized and tested for beta-lactamase inhibition. Selected active compounds have been assayed in antibacterial tests in combination with third-generation cephalosporins in order to investigate synergistic effects. Some results are shown in Table 2. Several of the compounds exhibit considerable efficacy in combination with ceftriaxone against *C. freundii* strains. The activity against *Pseudomonas* strains was weaker and the combinations were generally inactive against class A beta-lactamase-overproducing strains.

5.10 Conclusion

The new inhibitors exhibit potent synergy in antibacterial action when used in combination with existing antibiotics that are ineffective against the target organism because of the action of beta-lactamases [12]. As well as constituting a new direction in antibacterial chemotherapy, the results demonstrate the effectiveness of design based on understanding the details of enzymatic mechanism.

References

[1]   (a) Jarvis, W. R., and Martone, W. J., *J. Antimicrob. Chemother.* **29A**, 19 (1992); (b) Jones, R. N., Kehrberg, E. N., Erwin, M. E., and Anderson, S. C., *Diagn. Microbiol. Infect. Dis.* **19**, 203 (1994)
[2]   (a) Murray, B. E., *J. Infect. Dis.* **163**, 1185 (1991); (b) Travis, J., *Science* **264**, 360 (1994)
[3]   (a) Sanders, C. C., and Sanders, W. E. J., *Clin. Infect. Dis.* **15**, 824 (1992); (b) Davies, J., *Science* **264**, 375 (1994)
[4]   Poernuli, K. J., Rodigo, G., and Dornbusch, K., *J. Antimicrob. Chemother.* **34**, 943 (1994)
[5]   (a) Pratt, R. F., beta-Lactamase: inhibition. In: *The Chemistry of beta-lactams*, Page, M. I. (Ed.). Blackie Academic & Professional, London; 229–271 (1992); (b) Richter, H. G. F. et al., *J. Med. Chem.* **39**, 3712–3722 (1996)
[6]   Oefner, C., D'Arcy, A., Daly, J. J., Gubernator, K., Charnas, R. L., Heinze, I., Hubschwerlen, C., and Winkler, F. K., *Nature* **343**, 284–288 (1990)
[7]   (a) Herzberg, O., and Moult, J., *Science* **236**, 694 (1987); (b) Moews, P. C., Knox, J. R., Dideberg, O., Charlier, P., and Frere, J. M., *Proteins* **7**, 156–171 (1990)
[8]   Dubus, A., Ledent, P., Lamotte-Brasseur, J., and Frere, J.-M., *Proteins* **25**, 473–485 (1996)
[9]   Strynadka, N. C. J., Adachi, H., Jensen, S. E., Johus, K., Sielecki, A., Betzel, C., Sutoh, K., and James, J. N., *Nature* **359**, 700–705 (1992)
[10] (a) Dubus, A., Normark, S., Kania, M., and Page, M. G. P., *Biochemistry* **33**, 8577 (1994); (b) Dubus, A., Normark, S., Kania, M., and Page, M. G. P., *Biochemistry* **34**, 7757 (1995)
[11] Matsuda, K., Sanada, M., Nakagawa, S., Inoue, M., and Mitashashi, S., *Antimicrob. Agents Chemother.* **35**, 458 (1991)
[12] Angehrn, P., et al. in preparation.

[13] Page, M. G. P., *Biochem. J.* **295**, 295 (1993)

[14] Waley, S. G., *Biochem. J.* **279**, 87 (1991)

[15] Fersht, A., *Enzyme Structure and Mechanism*, 2. Edn., W. H. Freeman, New York, 296–299 (1984)

[16] (a) Bartlett, P. A., and Marlowe, C. K., *Science* **235**, 569–571 (1987); (b) Tronrud, D. E., Holden, H. M., and Matthews, B. W., *Science* **235**, 571–574 (1987)

[17] (a) Hansch, C., and Coats, E., *J. Pharm. Sci.* **59**, 731 (1970); (b) Yunger, L. M., and Cramer, D., *Mol. Pharmacol.* **20**, 602 (1981)

6 Inhibition of Sialidase

Neil R. Taylor

Abbreviations

Neu5Ac	*N*-acetyl-D-neuraminic acid
Neu5Ac2en	2-deoxy-2,3-didehydro-*N*-acetyl-D-neuraminic acid
4-amino-Neu5Ac2en	2-3-didehydro-2,4-dideoxy-4-amino-*N*-acetyl-D-neuraminic acid
4-guanidino-Neu5Ac2en	2-3-didehydro-2,4-dideoxy-4-guanidino-*N*-acetyl-D-neuraminic acid
MDCK	Madin-Darby canine kidney

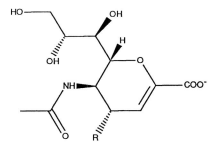

$R = OH$ (Neu5Ac2en)
$= NH_3^+$ (4-amino-Neu5Ac2en)
$= NHC(NH_2)_2^+$ (4-guanidino-Neu5Ac2en)

6.1 Introduction

The discovery of potent inhibitors of sialidase from influenza virus provides a classic example of a successful receptor-based drug design project and hints at the tremendous potential of combining protein crystallography and molecular modeling techniques in the field of medicinal chemistry. The basis of this discovery was the accurate determination of the three-dimensional structure of influenza virus sialidase complexed with substrate-mimics that bind in the enzyme active site. Molecular modeling of ligands in the binding site led directly to the synthesis of two compounds with binding affinities greater than any sialidase inhibitors known previously. One of these compounds has recently entered clinical trials as a possible treatment for influenza infection in humans.

6.2 Influenza: Disease and Virus

Influenza, or the 'flu, is an acute respiratory infection caused by a virus of the same name. Influenza epidemics continually have very high human costs in terms of excess mortality in high-risk groups, increased numbers of hospitalizations and medical treatment, and days lost at work and school. Symptoms of the disease include cough, soar throat, headache, fever, muscular aches, and prostration, and characteristically last 3–4 days. The severity of influenza is variable, determined both by the age, general health, and immunologic experience of the host and by the nature of the invading virus. Sometimes infection can occur completely without symptoms and in other instances the illness can be chronic and complications, such as pneumonia, can lead to a rapid death. In fact, influenza virus was responsible for the deadliest pandemic in history, during 1917–1918 it claimed the lives of more than twenty million people world-wide.

Influenza viruses belong to the orthomyxoviridae family. There are three types of influenza virus, named A, B and C (from the chronological order of their characterization). A multitude of subtypes have been characterized for type A virus. These have arisen from the antigenic cross-reactivities of the surface glycoproteins (see below). No subtypes have been identified for types B and C. Influenza type C virus is significantly different to the other two and in this work only influenza viruses A and B are discussed. Influenza virus particles are spherical in shape, approximately 100 nm in diameter. A schematic diagram of the virus is shown in Fig. 1.

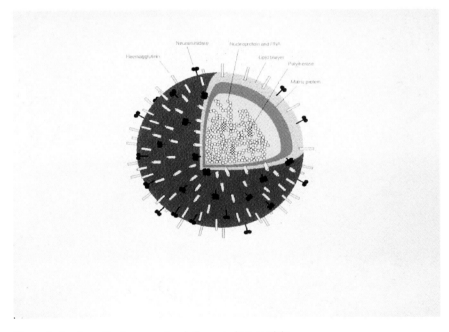

Figure 1. A schematic diagram of an influenza virus particle.

The external viral glycoproteins hemagglutinin and sialidase appear as spikes on the viral surface and number between 400 and 500 per particle [1]. Thirteen distinct antigenic subtypes have been identified for hemagglutinin, H1 to H13, and for sialidase nine subtypes have been characterized, N1 to N9. Only three combinations of these, H1N1, H2N2 and H3N2, have been isolated from human infections (the remaining subtypes were isolated from equine, swine and avian viruses). Hemagglutinin is believed to be important for both host cell recognition and binding and membrane fusion (reviewed in [2]). Sialidase was first thought to be required for both transporting the virus particle through mucin lining the respiratory tract [3–5] and for the release of newly synthesized virions from infected cells [6–8]. A very recent study, however, has shown that the enzyme is only important for virus release [9].

The most effective antiviral agents currently used to treat influenza are the drugs amantadine and its analog, rimantadine [10]. These drugs are useful prophylactics and treatments of infection by influenza A virus. They are understood to act by blocking M2, the ion channel protein in the viral envelope. Unfortunately, these drugs are of limited clinical use due to side effects associated with drug toxicity, the lack of influenza B activity, and the rapid emergence of resistant viral strains. Another drug that has shown to be effective in the treatment of in-

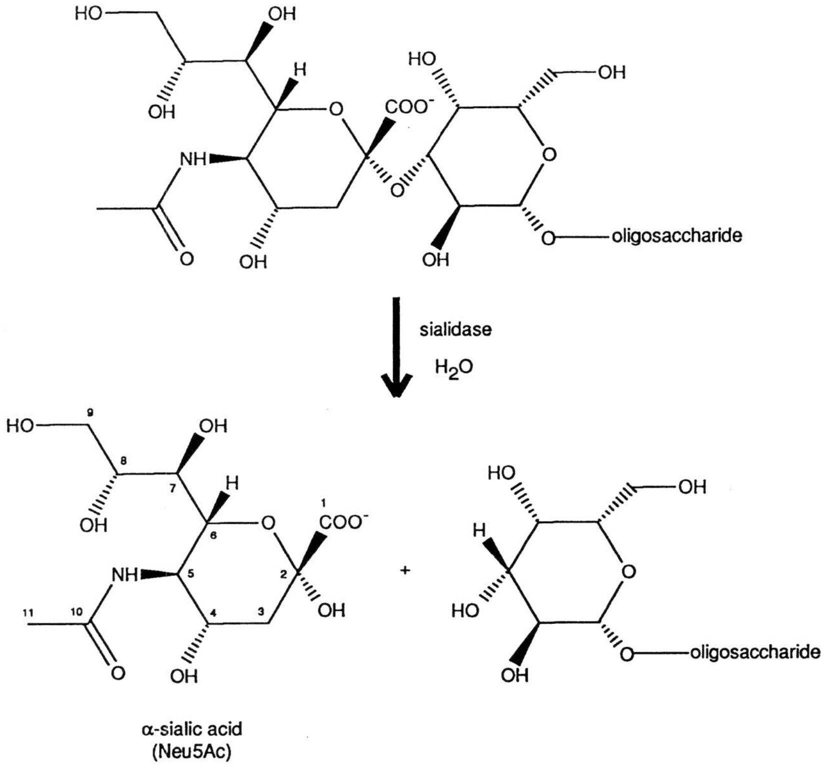

Scheme 1

fluenza is the synthetic triazole nucleoside ribavirin [10]. The mechanism of action of this broad-spectrum antiviral has not been proven, and it is currently undergoing trials for use in humans. In high-risk groups, namely health-care workers and the elderly, vaccinations are often administered to provide some protection from the currently circulating viruses [1]. Unfortunately the antigenic variability of the virus can only result in temporary protection.

6.3 Structure of Sialidase

Sialidase, EC 3.2.1.18 (also known as neuraminidase and acylneuraminyl hydrolase), is the name given to glycohydrolases which cleave terminal sialic acid (also known as neuraminic acid) from neighboring sugar units in glycoconjugates such as glycoproteins, glycolipids and polysaccharides (reviewed in [11–13]). Influenza virus sialidase shows greatest preference for *N*-acetyl-D-neuraminic acid, Neu5Ac, with a slight preference for an $\alpha2$–3 linkage to a galactose [14, 15]. Scheme 1 shows the general reaction. Sequence studies have shown that influenza virus sialidases from different serotypes can have up to 50% sequence variation [16]. Alignments however show that the active site residues are highly conserved between influenza A and B virus strains.

Treatment of influenza virus particles with protease leads to the release of the extraviral part of sialidase. The released protein is tetrameric, roughly rectangular prismatic in shape with dimensions $90 \times 90 \times 60$ Å, has a molecular weight of approximately 200 kDa, and it retains complete enzymatic and antigenic activity [17]. This 'head', which forms the bulk of the enzyme, is joined to the viral membrane by a short sequence containing many hydrophobic residues, the 'stalk', giving an overall mushroom-shaped topology. The tetramer has four-fold symmetry with the rotational axis normal to the membrane surface. Of the roughly 390 residues which comprise each monomer, approximately half of the residues make up β-sheet, in the form of six, anti-parallel, four-stranded sheets, arranged as if on the blades of a propeller, and the remaining half form a total of 24 loops and turns linking the β-strands into a single polypeptide segment [18–20]. Fig. 2 illustrates the backbone fold of the protein.

Two calcium ion binding sites have been identified in the protein, one on the four-fold symmetry axis near the outermost surface of the tetramer, and one approximately 8 Å from the active site. Calcium has been shown to increase the catalytic rate of the enzyme, although it is not absolutely necessary for activity [21, 22]. Four carbohydrate attachment sites have also

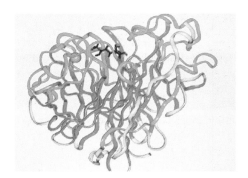

Figure 2. The monomeric subunit of sialidase showing the protein backbone, sialic acid (red) bound in the active site, and the calcium ion (yellow) adjacent to the active site. The loops that form the top surface of the protein, which includes the active site, are colored blue, the protein b-sheet component is colored green, and the loops that form the bottom of the protein and C-terminal segment, are colored gray.

been identified. The functional role of the oligosaccharide chains is not clear, although they have been found to be important in the crystallisation of the enzyme from influenza virus A [23].

The active site in the enzyme is easily recognizable in the crystal structure as a cavity, approximately 8 Å deep and 15 Å across, formed by loops on the outer-most surface of the protein. About 20 amino acids in and around the active site are strain-invariant. Many of the side chains that directly bind substrate are charged, and their net charge is approximately balanced. On the tetramer the active sites are located at a roughly 45° angle to the four-fold axis and are separated by approximately 45 Å.

Crystallographic methods have been used to probe the binding of monoclonal sialidase antibodies to the protein [24–26]. Crystal structures of antibody–antigen complexes exhibit the usual high degree of charge and surface complementarity observed at protein–protein interfaces. The complex of sialidase and the Fab of monoclonal antibody NC41 involves approximately 15 amino acid residues from five loops on the protein surface [26]. More than 600 Å^2 of protein surface area appears to be required for molecular recognition. Inhibition of the enzyme by the antibodies occurs by the steric blocking of the active site, thereby preventing binding of substrate. Mutation of a single amino acid residue on the surface of sialidase may possibly be sufficient to stop antibody binding to the antigen.

6.4 Mechanism of Catalysis

The mechanism of glycohydrolysis catalysis by sialidase has been investigated using a variety of approaches including site-directed mutagensis [27], kinetic isotope methods [28, 29], ^{1}H NMR [28], molecular modeling [28, 30], and X-ray crystallography [31, 32]. It has been proposed that the enzyme-catalyzed reaction occurs via an endocyclic sialosyl cation intermediate planar at C2 [28]. A solvent molecule, activated by the conserved residue Asp151, was suggested to donate a proton to the glycosidic oxygen atom. The resulting transition state intermediate has a positive charge in the pyranose ring which is stabilized by neighboring active site residues (see below). Subsequent attack by a water molecule and release of α-Neu5Ac was found to be the rate-limiting step. Overall, the reaction occurs with retention of configuration. Scheme 2 illustrates these reaction steps.

The zwitter-ionic transition state intermediate of the enzyme-catalyzed reaction is understood to be stabilized in the active site with the C2 carboxylate group held roughly coplanar with the ring atoms C2, C3 and O6 [28]. The active site arginine residues 118, 292, and 371 tightly bind the acid (see section 6.5), and it has been suggested that residues Asp151 and/or Glu277 stabilize the positive charge in the pyranose ring. A closely related proposal implicated the sidechain oxygen atom in Tyr406 in stabilizing the positively charged center, rather than a carboxylate group [33]. More recently, the results from crystal structure analyses on sialidase from influenza B soaked with various inhibitors and substrates suggested that cleavage of the glycosidic bond occurs prior to proton donation and that stabilization of the positive charge in the intermediate is achieved by multiple interactions with active site groups [32].

Despite the lack of a complete understanding of the mechanism of action of sialidase and a consensus on the ordering of the steps of hydrolysis, it is widely accepted that the interme-

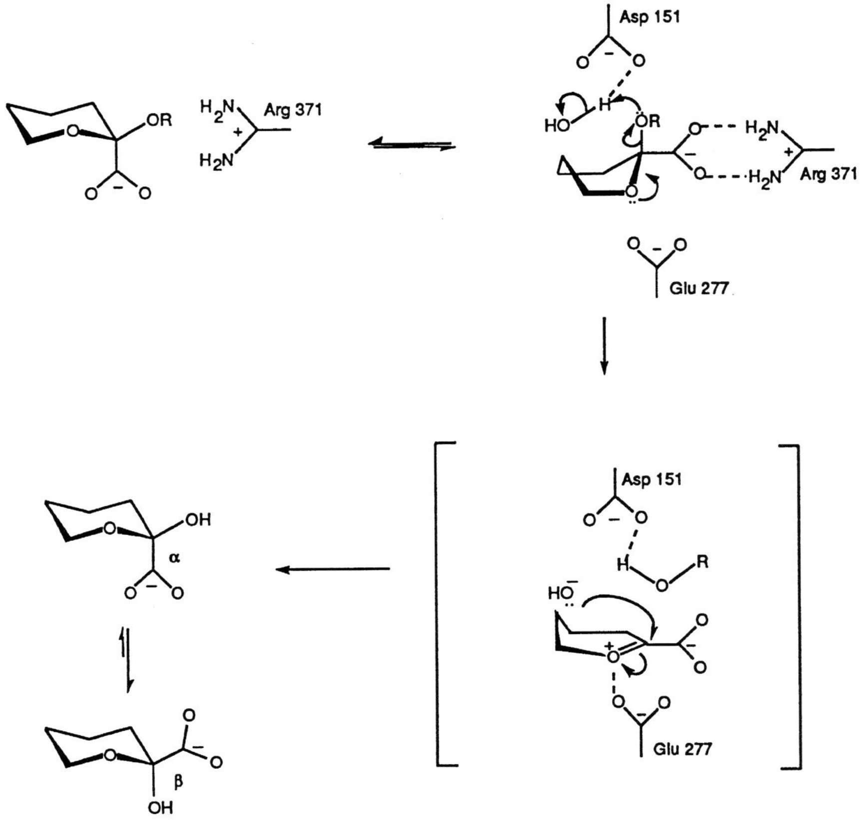

Scheme 2

diate in the reaction sequence is a Neu5Ac species roughly planar at C2. The compound 2-deoxy-2,3-didehydro-*N*-acetyl-D-neuraminic acid, Neu5Ac2en, first synthesized in 1969 [34], is a proposed transition state analog of the substrate. This compound has a K_i of 1 µM [35] which is an improvement of nearly two orders of magnitude over Neu5Ac. The drug design program described herein started with Neu5Ac2en as the lead compound and used information in the crystal structure of influenza virus sialidase to aid in the discovery of new synthetic inhibitors with enhanced binding and selectivity properties.

6.5 Binding of Substrate and Transition State Mimics

The mode of binding of Neu5Ac, the product of catalysis, to influenza virus sialidase was accurately determined using protein crystallography in the late 1980s [31]. Examination of the refined protein–ligand complex clearly shows that the enzyme binds the ligand not in its lowest energy $^2C_5$ solution conformation (*β*-Neu5Ac) but as the *α*-anomer with the pyranose ring in a distorted boat-like conformation. Overall, the functional groups of the ligand dock into the binding site with superb complementarity. The polar groups on Neu5Ac form favorable electrostatic and hydrogen-bonded interactions with side chains from nine residues and three buried water molecules in the active site of the enzyme (no backbone atoms make direct contact with the ligand). Furthermore, the hydrophobic portion in sialic acid makes favorable van der Waals interactions with a hydrophobic recognition site, comprised of two residues, in the binding cavity. The 11 residues interacting directly with the bound ligand are from very different positions along the polypeptide sequence; the fold of the protein brings these groups into close proximity. Comparison of the coordinates of the enzyme–ligand complex with the coordinates of the native enzyme shows that the active site side chains and buried water molecules do not significantly change position upon substrate binding. A stereo diagram of the key active site residues binding Neu5Ac is shown in Fig. 4 and the direct protein to ligand hydrogen bonding connectivities is illustrated schematically in Fig. 3.

Probably the most important sialidase–Neu5Ac interaction is the electrostatic force between the negatively charged carboxylate group of the ligand and the three positively charged arginines 118, 292 and 371. Residue Arg371 forms a planar salt-bridge to the acid through its two Nη atoms and residues Arg118 and Arg292, which are located on opposite sides of Arg371, each form a charge-charge based hydrogen bond to the acid through a Nη atom. The very tight binding of the carboxylic acid of Neu5Ac to this cluster of three arginine

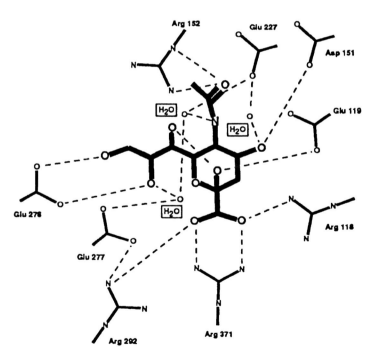

Figure 3. A schematic diagram showing the intermolecular hydrogen bonding network between influenza virus sialidase, including three buried water molecules, and bound Neu5Ac.

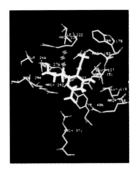 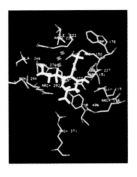

Figure 4. A stereo-view of the binding site of sialidase containing Neu5Ac (yellow).

residues, coupled with steric interactions between the pyranose ring and the floor of the active site, results in a distortion in the pyranose ring geometry upon binding. In the bound state, the sugar ring adopts a distorted boat-like conformation, the acid adopts an equatorial geometry, and the glycosidic oxygen atom ends up in an axial position. The importance of the carboxylate group in ligand binding to the enzyme has been clearly demonstrated for a variety of viral, bacterial, and mammalian sialidases; compounds without the acid group do not function as substrates or inhibitors [36–41].

Another very specific enzyme–substrate interaction involves the acetamido group which fits into a tight space, the C5 pocket, within the binding site. The carbonyl oxygen of this group forms hydrogen bonds with the Nε atom and one Nη atom from residue Arg152, which forms the roof of the pocket, the methyl group fits into a hydrophobic environment defined by residues Trp178 and Ile222, which form the back of the pocket, and the amide proton binds to a water molecule in the floor of the binding pocket. The water molecule is part of a hydrogen bonding network involving a number of water molecules and protein residues. A variety of C5 analogs of Neu5Ac2en were synthesized during the 1960s and 1970s and tested for enzyme activity. Replacement of the acetyl group by either a formyl or urea was found to reduce the K_i of Neu5Ac2en by more than two orders of magnitude and alkyl groups longer than a propyl showed no activity whatsoever [42–44]. Small gains were achieved in inhibitor potency by the stepwise replacement of protons of the methyl group by fluorine atoms [44]. It appears that the enzyme shows little tolerance to changes in this position. The gross binding mode of sialic acid is more or less defined by the salt bridge to Arg371, with the acetamido group locked into position.

The binding of the glycerol group to sialidase is another energetically important interaction. The two terminal hydroxyl groups, on C8 and C9, bind in a bidentate mode to Glu276, a residue that defines part of the floor of the binding cavity. The hydroxyl group on C8 also forms a strong hydrogen bond with a buried water molecule and the hydroxyl group on C9 interacts with a water molecule positioned above residue Arg224 which appears to be part of a hydrogen bonded network of waters. It has been shown chemically that the glycerol is very important for ligand binding, modifications to this functional group were found to be detrimental to the activity of the enzyme [45].

Two other hydroxyls in Neu5Ac form well-defined hydrogen bonds to the protein. The hydroxyl group on C2 interacts with residue Asp151. Despite being adjacent to arginine residues, Asp151 is not involved in a protein–protein salt-bridge. The hydroxyl group on C4 is accommodated in a pocket on the side of the binding cavity, the C4 pocket, fitting underneath the side chain of residue Asp151. It binds to a buried water molecule and forms a favorable electrostatic interaction with Glu119. The remaining hydroxyl group on C7 does not directly hydrogen bond with the protein, although results from a series of molecular mechanics and molecular dynamics calculations suggested that it may interact with Arg152 via a bridging water molecule (unpublished results). The ring oxygen does not appear to form a strong interaction with the protein, although some stability may arise from a partial charge–charge interaction involving Arg292.

Protein crystallography was also used to determine the mode of binding of the transition state analog Neu5Ac2en and sialidase [31]. The compound binds in the same manner as Neu5Ac, that is, the same set of enzyme–ligand hydrogen bonding interactions is observed for all functional groups they have in common. In solution, the sugar ring and acid group of Neu5Ac2en adopt a very similar conformation to the ring and acid group of bound Neu5Ac. The reduction of ring strain incurred at binding coupled with the mimicking of the transition state intermediate of sialidase accounts for increased stability of the complex over that involving Neu5Ac.

6.6 Structure-Based Inhibitor Design

The basic strategy undertaken for the discovery of potent influenza virus sialidase inhibitors was a combination of manual molecular graphics examination of the crystal structure containing bound Neu5Ac2en and application of the software package GRID which assists in *de novo* ligand design [46]. This program identifies favourable sites of interaction, or 'hot spots', between a protein binding site and small probe spheres which possess the properties of chemical functional groups. The program maps a binding cavity onto a three-dimensional grid, usually with a mesh size of around 1.0 Å. The energy of interaction between the protein and the probe is calculated at each grid point according to a molecular mechanics like energy scoring function. Using molecular graphics packages, the results of a GRID calculation can be readily displayed as contours of negative binding energy superimposed onto the bind-

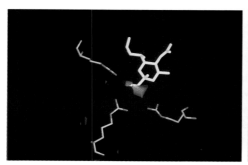
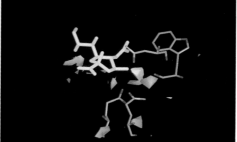

Figure 5. Calculated GRID contours for the active site of sialidase superimposed onto some of the amino acid residues and bound Neu5Ac (yellow) (a) Contours plotted at -10.0 kcal mol^{-1} for a carboxylate oxygen probe. (b) Contours plotted at -8.5 kcal mol^{-1} for an sp^3 amino nitrogen probe.

ing site. Fig. 5 illustrates two examples. Fig. 5(a) shows a negatively charged carboxylate probe sphere in the active site of sialidase and Fig. 5(b) shows a positively charged amine probe.

One of the most apparent hot spots in the active site of influenza sialidase was for an amino group in the pocket on the side of the binding cavity underneath the sidechain of residue Asp151 [47, 48]. In this position an amino group would form an energetically favorable ion-pair with the side chain of Glu119. The compound 2-3-didehydro-2,4-dideoxy-4-amino-*N*-acetyl-D-neuraminic acid, 4-amino-Neu5Ac2en, was synthesized [49] and found to be a potent inhibitor of influenza virus sialidase, with a K_i of 1×10^{-7} M [35]. Subsequent molecular graphics analyses suggested that the pocket on the side of the active site might just be large enough to accommodate the bulkier guanidino functional group. It was hypothesized that if the pocket was able to fit the guanidino substituent, its extended size and increased number of proton donor groups over the amino substituent might enable it to interact not only with the carboxylate on Glu119 but also the acid of Glu227. This compound 2-3-didehydro-2,4-dideoxy-4-guanidino-*N*-acetyl-D-neuraminic acid, 4-guanidino-Neu5Ac2en, was synthesized [49] and found to be much more potent than the 4-amino analog, with a K_i of between 10^{-10} and 10^{-11} M [47, 50].

6.7 Enzyme–Inhibitor Interactions

In order to investigate the intermolecular interactions involving the amine group a molecular mechanics energy minimization was done with 4-amino-Neu5Ac2en docked into the crystal structure of sialidase [30]. Fig. 6 illustrates part of the calculated structure. The favorable short-range electrostatic interaction between the acid group Glu119 and the basic substituent was calculated. The results also showed that the catalytic residue Asp151 was intimately involved in the tight binding of this inhibitor, and that a water molecule could occupy the C4 pocket and hydrogen bond to the amine, thereby filling the tetrahedral coordination sphere of the nitrogen atom.

The results of an energy minimization with 4-guanidino-Neu5Ac2en docked into the crystal structure of the enzyme are shown in Fig. 7. The predicted favorable electrostatic interactions involving the side chains of Glu119, which is to the side of the guanidine group, and of Glu227, which is below the guanidine, were observed in the low-energy structure. Furthermore, residue Asp151 was calculated to from a short hydrogen bond to the C4 substituent, as found for the amino group. Two other active site groups were observed to contribute to the binding of the 4-guanidino group, the backbone carbonyl of Trp178, which is positioned at the back of the C4 pocket and hydrogen bonds to the two terminal Nη atoms, and a buried water molecule, the one which hydrogen bonds to the Nη of the acetamido functional group. The fitting of the guanidinium moiety into the C4 pocket would be expected to displace the water molecule from this region, thereby providing an additional entropic contribution to the binding. A superimposition of the active sites for the complexes of sialidase with Neu5Ac2en and 4-guanidino-Neu5Ac2en demonstrated that there is very little change in the positions of the active site side chains when the C4 hydroxyl group is replaced by a guanidinium group. Finally, a very close agreement was obtained between the calculated binding mode for the inhibitor and the crystal structure of the complex [30].

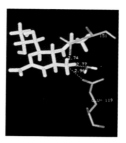

Figure 6. A stereodiagram showing the intermolecular hydrogen bonding interactions for the energy-minimized complex of 4-amino-Neu5Ac2en and sialidase in the vicinity of the amino moiety.

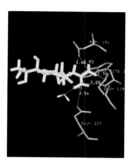
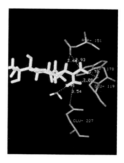

Figure 7. A stereodiagram showing the intermolecular hydrogen bonding interactions for the energy-minimized complex of 4-guanidino-Neu5Ac2en and sialidase in the vicinity of the guanidino moiety.

The observed increase in binding affinity on going from a hydroxyl group at the C4 position of Neu5Ac2en to an amine group, and then to a guanidino group, was qualitatively reproduced in a series of molecular modeling calculations [30, 51]. Table 1 lists the calculated energies. Values representing relative binding affinities were computed using two independent methods, firstly as the sum of molecular mechanics derived nonbonded interactions, and secondly as the change in total electrostatic energy, calculated using a continuum electrostatics approach. From these calculations the observed increase in inhibitor potency was traced to both specific hydrogen bonding interactions and long-range electrostatic forces, with the

Table 1. Inhibition of sialidase from influenza virus A/N2 by three transition state analogs and calculated relative binding enthalpies. Calculated results were obtained using molecular mechanics derived nonbonded interaction energies (MM energies) and a continuum electrostatics approach (CE energies).

Inhibitor	K_i (M)[a]	MM energies[b] (kcal mol^{-1})	CE energies[b] (kcal mol^{-1})
Neu5Ac2en	1×10^{-6}	-187	1
4-amino-Neu5Ac2en	1×10^{-7}	-211	-10
4-guanidino-Neu5Ac2en	3×10^{-11}	-223	-13

[a] [35, 50]
[b] [51]

electrostatic contribution between the basic groups and the negative residues Glu119 and Asp151 being the most important. In the case of 4-guanidino-Neu5Ac2en, the very large increase in binding affinity over the 4-amino analog suggests a significant entropic factor may be involved. This entropic contribution possibly arises from the displacement of the buried water molecule from the C4 pocket.

The inhibitor design study described above was performed using the active site of an N2 type sialidase extracted from an influenza A virus (A/Tokyo/3/67). Since that work was done, the crystal structures of three other examples of influenza virus sialidase have been reported, an N9 subtype from influenza virus A, A/Tern/Australia/G70c/75 [52], and the sialidases from two strains of influenza B virus, B/Beijing/1/87 [53] and B/Lee/40 [32]. The stereochemical arrangement of the conserved active site amino acid residues are very similar in all known influenza virus sialidases and most importantly, all have the pocket on the side of the binding cavity suitable for accommodating the basic amine and guanidine substituents in 4-amino-Neu5Ac2en and 4-guanidino-Neu5Ac2en, respectively. The structure of a non-viral sialidase has recently been reported and this is described in the next section.

An interesting feature of the interaction of 4-guanidino-Neu5Ac2en and influenza virus sialidase is that it exhibits slow-binding kinetics, with inhibition changing by several orders of magnitude over a 2-hour period [50, 54]. In the first study on this, slow-binding behavior was observed for sialidase from influenza A virus but not for enzyme from type B virus. In the latter report, sialidases from both influenza A and B viruses showed slow binding. The discrepancy in these results is unclear and is currently under investigation (Dr. M. von Itzstein, personal communication). One explanation for slow-binding kinetics that has been suggested involves the hindered release of the water molecule bound in the C4 pocket of the binding cavity in the enzyme [50]. None of the bacterial or mammalian sialidases that was studied showed slow-binding kinetics [50].

6.8 Inhibitor Potency, Efficacy and Selectivity

4-Guanidino-Neu5Ac2en has been shown to be an extremely potent inhibitor of the sialidases from a number of strains of influenza virus, usually with K_i values in the sub-nanomolar range (see Table 2). Inhibition of viral replication has been demonstrated in tissue culture, plaque formation in MDCK cells was reduced by 50% with concentrations subnanomolar for the majority of strains tested [47, 55]. These were significantly lower concentrations than the values obtained using the drugs amantadine, ramantadine and ribavirin [55]. In the results

Table 2. Inhibition of sialidases from different influenza virus sources by 4-guanidino-Neu5Ac2en

Influenza virus source	K_i (nM)[a]
N1 (from A/Brazil/11/78)	0.08
N2 (from A/Tokyo/3/67)	0.03
N2 (from reassortment X31)	0.2
N9 (from NSW/Whale)	0.2
B (from B/Beijing/1/87)	1.4
B (from B/Hong Kong/3/91)	0.7

[a] [47, 50, 54]

from animal studies using ferrets, 4-guanidino-Neu5Ac2en administered both before and during infection significantly reduced viral shedding in nasal washings and pyrexia in each animal tested [47]. The rationally designed drug was around three orders of magnitude more effective than amantadine when given intranasally. Cytotoxicity was not observed for the drug candidate with concentrations up to 10 nM for four cell lines [55]. Finally, and most importantly, the results from experimental infections in human adult volunteers show that the compound has efficacy in both a prophylactic and therapeutic mode [56]. Both the length and severity of influenza symptoms can be substantially reduced when 4-guanidino-Neu5Ac2en is administered intranasally before and during virus challenging.

Influenza viral particles are continually able to avoid the body's learned immune response mechanisms through the rapid mutation of its antigens. The virus is similarly able to respond to pressure from the drugs amantadine and rimantadine [57]. The possibility of drug-resistant strains arising in response to 4-guanidino-Neu5Ac2en is therefore an important consideration and is currently under investigation. Some preliminary results have been favorable. It was demonstrated that passage history was not significant in affecting inhibition by 4-guanidino-Neu5Ac2en [55] and resistant isolates were not found in laboratory studies using conditions which give rise to amantadine resistance [47].

Inhibition studies have shown that 4-amino-Neu5Ac2en and, in particular, 4-guanidino-Neu5Ac2en, have much higher affinities for sialidase from influenza virus than for sialidases from other viral sources, and far greater affinities than for either bacterial or mammalian sialidases [35] (see Table 3). In a recent study, a series of synthetic modifications were made to the C4 substituent of Neu5Ac2en and all compounds were found to be weaker inhibitors of bacterial and mammalian sialidases than the unmodified starting molecule [35]. The recently reported three-dimensional structure of a bacterial sialidase, from *Salmonella typhimurium* LT2, clearly shows that the active site lacks the acidic residues equivalent to Glu119 and Glu227 in the C4 pocket [58]. The active site is thus less suited to accommodate a charged C4 substituent on the Neu5Ac2en template.

6.9 Summary

The computer-aided design of potent inhibitors of influenza virus sialidase based on a crystal structure of the enzyme has been described. A combination of manual molecular graphics analysis and the *de novo* ligand design tool GRID was used to identify modifications to the

Table 3. Inhibition of sialidases from different viral, bacterial, and mammalian sources by 4-guanidino-Neu5Ac2en

Enzyme source	K_i (μM)[a]
Parainfluenza virus	800
Clostridium perfringens	> 100
Vibrio cholerae	60
Arthrobacter ureafaciens	> 10 000
Sheep liver	300

[a] [35]

transition state analog, Neu5Ac2en, that would increase its binding affinity to the enzyme. Molecular modeling predicted that replacement of a hydroxyl group on the lead compound by a basic substituent should produce favorable electrostatic interactions between enzyme and ligand. The compounds 4-amino-Neu5Ac2en and 4-guanidino-Neu5Ac2en were synthesized and found to be potent inhibitors of influenza virus sialidase. The 4-guanidino compound in particular has shown excellent selectivity and *in vivo* efficacy and therefore represents a very promising new drug candidate for influenza virus infection in humans. The approach described in this report for designing inhibitors based on the known three-dimensional structure of a protein binding site can readily be applied to any biomolecular system.

Acknowledgements

I thank Dr. Peter McMeekin for providing the data for the GRID maps shown here. Thanks are also extended to Dr. Mike Hann, Dr. Mark von Itzstein and Dr. Peter Cherry for their assistance with the final manuscript.

References

[1] Kilbourne, E. D. *Influenza.* Plenum: New York 1987
[2] Klenk, H.-D., and Rott, R., *Adv. Virus Res.* **34**, 247–281 (1988)
[3] Burnet, F. M., McCrea, J. F., and Anderson, S. G., *Nature* **160**, 404–405 (1947)
[4] Burnet, F. M., *Aust. J. Exp. Biol. Med. Sci.* **26**, 381–387 (1948)
[5] Colman, P. M., and Ward, C. W., *Curr Top. Microbiol. Immun.* **114**, 117–255 (1985)
[6] Palese, P., Tobita, K., Ueda, M., and Compans, R. W., *Virology* **61**, 397–410 (1974)
[7] Palese, P., and Compans, R. W., *J. Gen. Virol.* **33**, 159–163 (1976)
[8] Griffin, J. A., and Compans, R. W., *J. Exp. Med.* **150**, 379–391 (1979)
[9] Liu, C., Eichelberger, M. C., Compans, R. W., and Air, G. M., *J. Virol.* **69**, 1099–1106 (1995)
[10] von Itzstein, M., Barry, J. B., and Chong, A. K. J., *Curr. Opin. Therap.* **3**, 1755–1762 (1993)
[11] Drzeniek, R., *Curr. Top. Microbiol. Immun.* **59**, 35–74 (1972)
[12] Air, G. M., and Laver, W.G., *Proteins* **6**, 341–356 (1989)
[13] Colman, P. M. Neuraminidase: Enzyme and Antigen. In: *The Influenza Viruses*, Krug, R. M. (Ed.). Plenum: New York; 175–218 (1989)
[14] Corfield, A. P., Wember, M., Schauer, R., and Rott, R., *Eur. J. Biochem.* **124**, 521–525, 1982
[15] Corfield, A. P., Higa, H., Paulson, J. C., and Schauer, R., B*iochim. Biophys. Acta* **744**, 121–126 (1983)
[16] Colman, P. M. The Structure and Function of Neuraminidase. In: *Peptide and Protein Reviews*, Vol. 4, Hearn, M. T. (Ed.). Marcel Dekker: New York; 215–255 (1984)
[17] Laver, W. G., *Virology* **86**, 78–87 (1978)
[18] Varghese, J. N., Laver, W. G., and Colman, P. M., *Nature* **303**, 35–40 (1983)
[19] Colman, P. M., and Varghese, J. N., *Nature* **303**, 41–44 (1983)
[20] Varghese, J. N., and Colman, P. M., *J. Mol. Biol.* **221**, 473–486 (1991)
[21] Chong, A. K. J. *Influenza Virus Sialidase: A Mechanistic Study.* Masters Thesis, Monash University, Melbourne 1990
[22] Chong, A. K. J., Pegg, M. S., and von Itzstein, M., *Biochim. Biophys. Acta* **1077**, 65–71 (1991)
[23] Colman, P. M., *Prot. Sci.* **3**, 1687–1696 (1994)
[24] Colman, P. M., Laver, W. G., Varghese, J. N., Baker, A. T., Tulloch, P. A., Air, G. M., and Webster, R. G., *Nature* **326**, 358–363 (1987)
[25] Tulip, W. R., Varghese, J. N., Webster, R. G., Laver, W. G., and Colman, P. M., *J. Mol. Biol.* **227**, 149–159 (1992)
[26] Malby, R. L., Tulip, W. R., Harley, V. R., McKimm-Breschkin, J. L., Laver, W. G., Webster, R. G., and Colman, P. M., *Structure* **2**, 733–746 (1994)

[27] Lentz, M. R., Webster, R. G., and Air, G. M., *Biochemistry* **26**, 5351–5358 (1987)

[28] Chong, A. K. J., Pegg, M. S., Taylor, N. R., and von Itzstein, M., *Eur. J. Biochem.* **207**, 335–343 (1992)

[29] Guo, X., Laver, W. G., Vimr, E., and Sinnott, M. L., *J. Am. Chem. Soc.* **116**, 5572–5578 (1994)

[30] Taylor, N. R., and von Itzstein, M., *J. Med. Chem.* **37**, 616–624 (1994)

[31] Varghese, J. N., McKimm-Breschkin, J. L., Caldwell, J. B., Kortt, A. A., and Colman, P. M., *Proteins* **14**, 327–332 (1992)

[32] Janakiraman, M. N., White, C. L., Laver, W. G., Air, G. M., and Luo, M., *Biochemistry* **33**, 8172–8179 (1994)

[33] Burmeister, W. P., Henrissat, B., Bosso, C., Cusack, S., and Ruigrok, R. W. H., *Structure* **1**, 19–26 (1993)

[34] Meindl, P., and Tuppy, H., *Monatsh. Chem.* **100**, 1295–1306 (1969)

[35] Holzer, C. T., von Itzstein, M., Jin, B., Pegg, M. S., Stewart, W. P., and Wu, W.-Y., *Glycoconjug. J.* **10**, 40–44 (1993)

[36] Brossmer, R., and Holmquist, L., *Hoppe-Seyler's Z. Physiol. Chem.* **352**, 1715–1719 (1971)

[37] Brossmer, R., Bürk, G., Eschenfelder, V., Holmquist, L., Jäckl, R., Neumann, B., and Rose, U., *Behring Inst. Mitt.* **55**, 119–123 (1974)

[38] Miller, C. A., Wang, P., and Flashner, M., *Biochem. Biophys. Res. Commun.* **83**, 1479–1487 (1978)

[39] Kumar, V., Kessler, J., Scott, M. E., Patwardham, B. H., Tanenbaum, S. W., and Flashner, M., *Carbohydr. Res.* **94**, 123–130 (1981)

[40] Kumar, V., Tanenbaum, S. W., and Flashner, M., *Carbohydr. Res.* **103**, 281–285 (1982)

[41] Eschenfelder, V., Brossmer, R., and Wachter, M., *Hoppe-Seyler's Z. Physiol. Chem.* **364**, 1411–1417 (1983)

[42] Meindl, P., and Tuppy, H., *Monatsh. Chem.* **97**, 990–999 (1969)

[43] Meindl, P., and Tuppy, H., *Monatsh. Chem.* **97**, 1628–1647 (1969)

[44] Meindl, P., Bodo, G., Palese, P., Schulman, J., and Tuppy, H., *Virology* **58**, 457–463 (1974)

[45] Suttajit, M., and Winzler, R. J., *J. Biol. Chem.* **246**, 3398–3404 (1971)

[46] Goodford, P. J., *J. Med. Chem.* **28**, 849–857 (1985)

[47] von Itzstein, M., Wu, W.-Y., Kok, G. B., Pegg, M. S., Dyason, J. C., Jin, B., Van Phan, T., Smythe, M. L., White, H. F., Oliver, S. W., Colman, P. M., Varghese, J. N., Ryan, D. M., Woods, J. M., Bethell, R. C., Hotham, V. J., Cameron, J. M., and Penn, C. R., *Nature* **363**, 418–423 (1993)

[48] von Itzstein, M., Dyason, J. C., White, H. F., and Oliver, S. W., *J. Med. Chem.* **39**, 388–391 (1996)

[49] von Itzstein, M., Wu, W.-Y., and Jin, B., *Carbohyd. Res.* **259**, 301–305 (1994)

[50] Pegg, M. S., and von Itzstein, M., *Biochem. Mol. Biol. Int.* **32**, 851–858 (1994)

[51] Taylor, N. R., and von Itzstein, M., *J. Comp.-Aided Mol. Des.* **10**, 233–246 (1996)

[52] Tulip, W. R., Varghese, J. N., Baker, A. T., van Donkelaar, A. Laver, W. G., Webster, R. G., and Colman, P. M., *J. Mol. Biol.* **221**, 487–497 (1991)

[53] Burmeister, W. P., Ruigrok, R. W. H., and Cusack, S., *EMBO J.* **11**, 49–56 (1991)

[54] Hart, J. H., and Bethell, R.C., *Biochem. Mol. Biol. Int.* **36**, 695–702 (1995)

[55] Woods, J. M., Bethell, R. C., Coates, J. A. V., Healy, N., Hiscox, S. A., Pearson, B. A., Ryan, D. M., Ticehurst, J., Tilling, J., Walcott, S. M., and Penn, C. R., *Antimicrob. Agents Chemother.* **37**, 1473–1479 (1993)

[56] Hayden, F. G., Lobo, M., Esinhart, J., and Hussey, E., *Efficacy of 4-guanidino Neu5Ac2en in Experimental Human Influenza A Virus Infection.* 34th Interscience Conference on Antimicrobial Agents and Chemotherapy. Orlando, 4–7 October (1994). American Society for Microbiology, Washington DC 1994, p. 190

[57] Hayden, F. G., Sperber, S. J., Belshe, R. B., Clover, R. D., Hay, A. J., and Pyke, S., *Antimicrob. Agents Chemother.* **35**, 1741–1747 (1991)

[58] Crennell, S. J., Garman, E. F., Laver, W. G., Vimr, E. R., and Taylor, G. L., *Proc. Natl Acad. Sci. USA* **90**, 9852–9856 (1993)

7 Rational Design of Inhibitors of HIV-1 Reverse Transcriptase

Wolfgang Schäfer

Abbreviations

RT   Reverse Transcriptase
HIV   Human Immunodeficiency Virus
AIDS   Acquired Immunodeficiency Syndrome
NNRTI   Non-Nucleoside-RT-Inhibitor
MNDO   Modified Neglect of Diatomic Differential Overlap

7.1 Introduction

Reverse transcriptase (RT) is a key enzyme in the replication of human immunodeficiency virus and is therefore – besides the HIV-protease – a main target in the investigation of drugs against AIDS [1]. Today, two main groups of inhibitors of RT are known. First, the nucleoside analogs such as AZT (azidothymidine), which simulate the natural substrate of the enzyme; though their chain-terminating property could lead to toxic side effects. The second group, the so-called Non-Nucleoside-RT-Inhibitors (NNRTI), seem to offer an alternative means of therapy.

When we started our studies on this type of compound the most prominent examples of this class were TIBO (**1**) [2] and nevirapine (**2**) [3]. In our company, a high throughput screening of 30 000 compounds led to isoindolinon (**3**) whose activity against RT is in the same order of magnitude as that of **1** and **2** [4]. The inhibition data for this compound as well as for all other molecules referred to in this article are shown collectively in Table 1.

Table 1. In vitro activity of HIV-RT inhibitors. Values refer to the active enantiomers whose absolute configurations were determined by X-ray crystallography, except for **2** which is achiral.

Compound	IC$_{50}$ [µM]
1	0.16
2	0.43
3	0.28
4	8.70
5	0.15
6	0.03
7	0.83

The question arose whether there was a possibility of deriving improvements of our compound or to propose new lead compounds from comparison of the known molecules and some common properties. Since an X-ray structure of HIV-RT was not available at that time, we had to base our study on a comparison of properties of the inhibitor molecules alone. If such an approach includes molecules with conformational flexibility – as in **1** – a procedure of this type is easily subject to overinterpretation and must therefore be supported by experimental facts. This means that assumptions derived from a theoretical consideration should be checked by real test molecules before any conclusions can be drawn.

At first sight, a two-dimensional comparison of **1**, **2** and **3** yields no relationship between the molecules. Some similarities can be found in a three-dimensional approach: each system contains two π systems arranged in a butterfly-like orientation, with an additional lipophilic region between them, and a carbonyl or thiocarbonyl group. Therefore it seems reasonable to base a comparison on the structure of these molecules.

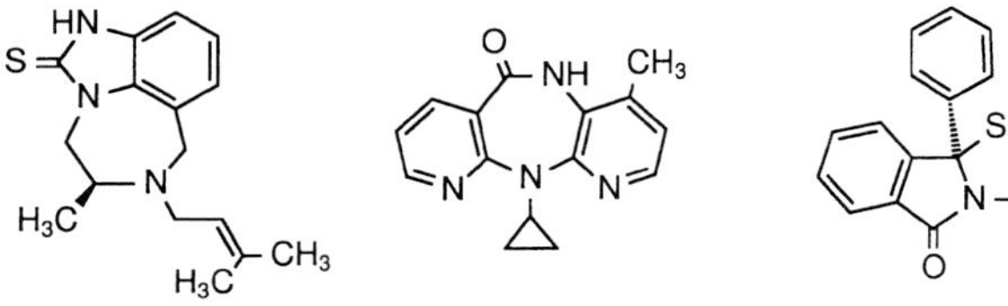

Scheme 1

The main structural features of compound **1** are the seven-membered ring and the orientation of the dimethylallyl group. We believe that it is not possible by calculation accurately to predict the exact conformation of the side chain; the usefulness of a crystal structure is also limited concerning this flexible group. The seven-membered ring contains three planar atoms and has some similarity with cyclohexene; we expect therefore the existence of two conformers that are related to the two twist forms of cyclohexene. A conformational analysis using MNDO [5] supports this assumption. For our model we have used the conformer with the lowest MNDO heat of formation, which is also in agreement with a published X-ray structure of a chloro derivative of TIBO [6].

Compound **2** was simply optimized by MNDO, since there are only two possible conformations (cyclopropyl group *syn* or *anti* to the seven-membered ring) and the mirror image of each of them. It is clear that any selection of the 'enantiomer' to be used in the comparison is arbitrary and may cause problems.

The third compound (**3**) contains only two degrees of freedom, namely the pseudorotation of the thiazolidine ring and the torsion of the exocyclic phenyl ring. An X-ray analysis of the active enantiomer was performed and yielded essentially the same structure as predicted by a conformational analysis with MNDO.

7.2 Building a Model

Starting from these structures, the electrostatic potentials have been calculated using MNDO point charges. The results are displayed in Fig. 1. It is striking that the structures exhibit a extremely similar spatial distribution of this property, although the molecules show only few similarities in a two-dimensional aspect. It is therefore reasonable to suppose that the three compounds could be recognized by an enzyme in a similar manner.

We have now moved these maps in space in order to obtain an optimal superimposition, disregarding a match of the underlying molecules. As can be seen in Fig. 2, this superimposition leads to an orientation of the underlying structures that does not represent an atom-by-atom match.

From this result the following model was derived (see Fig. 3): An aromatic system and an additional π-system should be arranged in a roof-like orientation, separated by a lipophilic site; furthermore it should contain a (thio-)carbonyl group in an arrangement as depicted in the figure, and a methyl group.

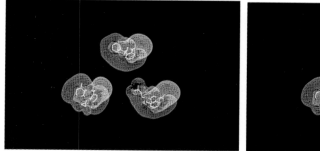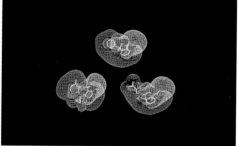

Figure 1. Electrostatic isopotential maps as derived from a point charge model using MNDO charges: -5 kcal mol^{-1}, cyan; $+5$ kcal mol^{-1}, purple. Hydrogen atoms omitted. Top, **3**; bottom left, **2**; bottom right, **1**.

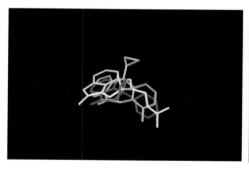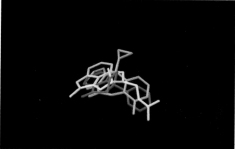

Figure 2. Orientation of **1** (yellow, MNDO structure), **2** (purple, MNDO structure), and **3** (blue, crystal structure) as result of the superposition of the electrostatic potential maps.

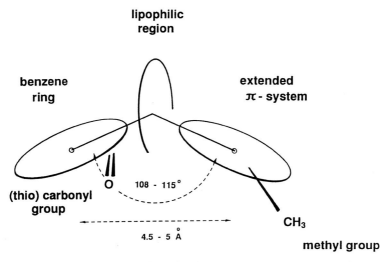

Figure 3. The common three-dimensional model derived from the comparison of **1**, **2** and **3**.

7.3 Test of the Model

In order to decide whether this model is correct, we have been searching the ACD database of commercially available compounds by use of two-dimensional search techniques with visual inspection of the hits [6]. Possible candidates were optimized and compared to the model. By this method, one single compound (**4**) was found that fits into the model. The enantiomers of **4** were separated, and one enantiomer showed some *in vitro* inhibitory activity (see Table 1); the other was inactive.

4	**5**	**Scheme 2**

7.4 Optimization of Compound 3

Superimposition of **3** and **4**, as shown in Fig. 4, showed that the phenyl rings on the left side occupied different positions in space. If both compounds were accepted by the enzyme, it followed that both positions should have been possible simultaneously. It was therefore imme-

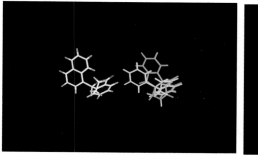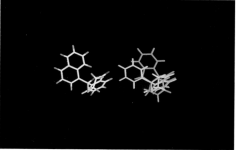

Figure 4. Right: superimposition of the crystal structures of **3** (cyan) and **4** (yellow). Left: one of the two low energy MNDO conformations of the proposed naphthyl compound **5**.

diately suggested by the figure that the naphthalene compound **5** should be synthesized and tested. The activity of this compound was indeed shown to have increased by a factor of 2 when compared to the lead compound **3** (Table 1).

None of the compounds mentioned so far contained the methyl group indicated in the model. From the comparison of **3** with nevirapine, a possible position for a methyl group in the exocyclic ring could be *ortho* or *meta* with respect to the bridgehead; inspection of the side chain of **1**, however – especially in a stereo representation (Fig. 2) – immediately suggests a close relationship to a *meta* substituted phenyl ring. Therefore the methyl group was introduced into **3** in a *meta* position, the resulting active enantiomer (**6**) showing an IC_{50} of 0.03 μM. Thus, it was possible by use of the model to increase the activity of the lead compound by a factor of 10.

6   **7**   **Scheme 3**

7.5 Design of a New Inhibitor

Apart from the variation of the starting compound **3**, which led to a number of highly potent inhibitors [4], we have also tried to design molecules that fit into the model but belong to different chemical classes and might serve as new leads. Structures and potential distributions of various molecules were calculated and compared to the model. Finally, compound **7** was found that fulfills both the geometric and the electronic requirements. As in the case of compounds **1**, **2** and **3**, there seem to be only few similarities between **3** and **7** as long as only the two-dimensional structures are considered. In a three-dimensional superimposition, however, the structures are almost identical (Fig. 5). The same holds for the isopotential map, which is included in Fig. 5. The compound was synthesized and tested; the activity is given in Table 7.1.

Although in this case the activity of the designed compound was poorer than that of the original lead, this molecule is in our opinion a nice example of rational drug design, because it was derived solely by modeling methods without knowledge of the structure of the biological target.

We would like to note that besides our modeling approach a large number of compounds was synthesized according to classical chemical lead optimization [4], though no major improvement was achieved. Furthermore we would like to mention that our compounds are not ideal as they exhibit both inadequate bioavailability and resistance problems. We trust, however, that some of these problems may be overcome by future studies.

7.6 Guidelines and Conclusions

Based on our experimental findings, the following guidelines should be followed regarding these types of studies.

1. Ensure that the inhibition constants of the test systems refer to a single compound. In the case of chiral compounds, separate the enantiomers. Check the activity of both enantiomers. Determine the absolute configuration of the active enantiomer.
2. Determine the structures of active *and* inactive compounds experimentally, preferably by X-ray crystallography.
3. Design compounds that make it possible to distinguish whether your model is correct or not.
4. If you are forced to design compounds by superposition of known structures, do not simply use an atom-by-atom match, but compare the properties of the compounds.

A number of crystal structures of RT complexed with several non-nucleoside inhibitors have recently been reported [7]. These structures also contain complexes with **1** and **2**. All

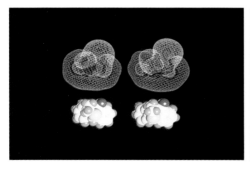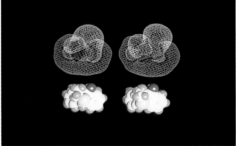

Figure 5. Comparison of the crystal structures of **7** (left) and **3** (right). Top: electrostatic potentials as described in the legend of Figure 1. Bottom: space filling models.

published inhibitor structures appear to have an overall shape that is in agreement with the model. It seems that the conformation of **1** is very close to the one we have been using. Compound **2**, however, seems to exhibit a different conformation. Obviously, the authors report the mirror image of **2** compared to our conformation. In this case this compound would bind differently to the active site than predicted by our model, although the overall shape of these enantiomeric conformations is very similar and might not be conclusively distinguishable at the resolution of the X-ray structure of the complex. Therefore, a final validation of the model presented here has to be postponed until complex structures of the above-mentioned complexes are available. We have also initiated co-crystallization of compounds **5** and **7** with RT with which we hope to obtain the final answer to this question.

References

[1] (a) Fauci, A. S., *Science* **239**, 617–622 (1988); (b) Mitsuya, H., Yarchoan, R., and Broder, S., *Science* **249**, 1533–1544 (1990); (c) Sarin, P. S., *Annu. Rev. Pharmacol.* **28**, 411–428 (1988)

[2] (a) Kukla, M. J., Breslin, H. J., Pauwels, R., Fedde, C. L., Miranda, M., Scott, M. K., Sherill, R. G., Raeymaekers, A., Van Gelder, J., Andries, K., Janssen, M.A.C., De Clercq, E., and Janssen, P.A.J., *J. Med. Chem.* **34**, 746–751 (1991); (b) Pauwels, R., Andries, K., Desmyter, J., Schols, D., Kukla, M. J., Breslin, H. J., Raeymaeckers, A., Van Gelder, J., Woestenborghs, R., Heykants, J., Schellekens, K., Janssen, M. A. C., De Clercq, E., and Janssen, P. A. J., *Nature* **343**, 470–474 (1990)

[3] (a) Cohen, K. A., Hopkins, J., Ingraham, R. H., Pargellis, C., Wu, J. C., Palladino, D. E. H., Kinkade, P., Warren, T. C., Rogers, S., Adams, J., Farina, P. R., and Grob, P. M., *J. Biol. Chem.* **266**, 14670–14674 (1991); (b) Hargrave, K. D., Proudfoot, J. R., Grozinger, K. G., Cullen, E., Kapadia, S. R., Patel, U. R., Fuchs, V. U., Mauldin, S. C., Vitous, J., Behnke, M. L., Klunder, J. M., Pal, K., Skiles, J. W., McNeil, D. W., Rose, J. M., Chow, G. C., Skoog, M. T., Wu, J. C., Schmidt, G., Engel, W. W., Eberlein, W. G., Saboe, T. D., Campbell, S. J., Rosenthal, A. S., and Adams, J., *J. Med. Chem.* **34**, 2231–2241 (1991)

[4] Mertens, A., Zilch, H., König, B., Schäfer, W., Poll, T., Kampe, W., Seidel, H., Leser, U., and Leinert, H., *J. Med. Chem.* **36**, 2526–2535 (1993)

[5] (a) Dewar, M. J. S., and Thiel, W., *J. Am. Chem. Soc.* **99**, 4899–4907 (1977); (b) Bischof, P., and Friedrich, G., *J. Comput. Chem.* **3**, 486–494 (1982)

[6] ACD database and program MACCS supplied by Molecular Design Ltd.; San Leandro, CA. 3-D-database programs were not available to us

[7] (a) Kohlstaedt, L. A., Wang, J., Friedman, J. M., Rice, P. A., and Steitz, T. A., *Science* **256**, 1783–1790 (1992); (b) Ding, J., Das, K., Moereels, H., Koymans, L., Andries, K., Janssen, P. A. J., Hughes, S. H., and Arnold, E., *Nature Struct. Biol.* **2**, 407–415 (1995); (c) Ren, J., Esnouf, R., Garman, E., Somers, D., Ross, Kirby, I., Keeling, J., Darby, G., Jones, Y., Stuart, D., and Stammers, D., *Nature Struct. Biol.* **2**, 293–302 (1995); (d) Esnouf, R., Ren, J., Ross, C., Jones, Y., Stammers, D., and Stuart, D., *Nature Struct. Biol.* **2**, 303–308 (1995)

8 New Computational Approaches to Predict Protein–Ligand Interactions

Hans-Joachim Böhm

8.1 Introduction

The previous chapters of this book have clearly demonstrated the enormous value of 3D protein structures in ligand design. In this structure-based ligand design [1–9], molecular modeling tools serve to visualize and analyze the structural information and to explore new possibilities through model building. 3D structures of protein–ligand complexes can be used to identify the essential interactions in these structures and to search for additional binding sites which are not used by the previously known ligands. Possible binding sites may be positions where hydrogen bonds can be formed with the enzyme, or hydrophobic pockets in the enzyme structure which can be occupied by lipophilic groups. So far, the critical aspect of the selection of the structure to be evaluated (and then possibly synthesized and tested) was left to the creativity of the medicinal chemist. The efficacy of the computer-aided drug design process could be greatly improved by additional tools which propose new molecules as possible ligands by an automatic or semi-automatic procedure [10, 11]. A number of computational tools have recently been described to select putative ligands and to predict their interactions with the protein. In general, these novel tools for drug design can be divided into the following major categories:

1. Analysis of the protein structure
2. Ligand docking and 3D database searching
3. *De novo* ligand design
4. Assessment of the ligand binding affinity

The first step in the search for a new ligand is the analysis of the 3D structure of the protein. The surface of the protein can be displayed together with certain properties such as the lipophilicity [12, 13] or the electrostatic potential [14]. Computational methods have been described to detect clefts or cavities in the protein [15–17]. Other approaches are available to predict favorable binding sites for probes or small molecules such as a water molecule or a carbonyl group [18–20]. One of the most frequently used approaches is to calculate interaction energies with different probes on a grid spanning the binding site. Such grids can be displayed and then subsequently be used as a guide in the design of new ligands. The program MCSS developed by Miranker and Karplus [21] combines the analysis of the protein binding site with a placement of small functional groups in an energetically favorable orientation. MCSS uses a molecular mechanics force field for the placement of small ligands.

The next step is the selection of a putative ligand. Basically, one can either search 3D databases of known small molecules or one can attempt a *de novo* design. The latter approach tries to construct a new molecule completely or in part from scratch. The possible approaches are sketched in Fig. 1.

The computationally most direct way to find a new ligand is to search a database containing 3D structures [11, 22, 23]. The pioneering program in this field is DOCK developed by Kuntz and coworkers (Fig. 1, left) [24]. The basic idea of DOCK is to search a 3D database for possible ligands based on shape complementarity between protein and ligand. Recently, the ability to account for electronic complementarity was also incorporated by means of a molecular mechanics force field [25, 26]. A number of successful applications of DOCK have been reported [27–29]. Further computer programs for 3D docking have been described [30, 31]. CLIX [30] uses the GRID force field [18, 19] in the docking of putative ligands. The program FLOG developed by Miller et al. [31] accounts for ligand flexibility by including up to 25 conformers of each structure in the database. The program LUDI described in the following section can also be used for 3D database searching [32].

A number of computer programs have recently been proposed that attempt to design automatically new ligands for a given protein structure [33–56]. Several reviews have been published [33–35]. Most programs try to assemble novel molecules from pieces. These pieces are either atoms [35–38] or larger, chemically reasonable fragments [39–41, 55, 56]. Atom-based *de novo* design programs are LEGEND [36], GROWMOL [37] and MCDNGL [38]. Fragment-based programs are GROW [39], NEWLEAD [40], GROUPBUILD [41], HOOK [42], TORSION [44] and LUDI [55, 56]. Both approaches have advantages and disadvantages. The use of single atoms as building blocks can generate the largest possible diversity of chemical structures. However, a large structural diversity can also be obtained with a fragment-based approach by using a large variety of different fragments. For example, in LUDI about 30 000 small rigid molecules were extracted from the fine chemicals directory (FCD) and are used as fragments [32]. A potential advantage of the fragment-based approach as compared to the atom-based approach concerns the synthetic accessibility of the generated structures. The use of fragments offers the advantage that chemical knowledge can be built into the fragment connection step. For example, amino acids can be used as building blocks to construct peptides [39]. The extension to other simple chemical reactions, e.g., the ether formation, is straightforward. In contrast, this control of synthetic accessibility is difficult to achieve for atom-by-atom build-up programs. Therefore, the latter approach requires that the synthetic accessibility is checked at the very end of the design cycle. This is a much more complex task than to check whether the formation of a particular bond is synthetically feasible.

Two major strategies exist for the fragment-based approach to *de novo* ligand design. The first possibility is to place several fragments independently (or take them from a known ligand structure) and then to search for suitable templates that connect these fragments into one molecule (Fig. 1, right). The advantage of this approach is that the individual fragments are placed without any bond constraints and are likely to be at their optimal positions. Furthermore, this strategy has the desired tendency to generate rigid structures. A possible disadvantage is that it may be difficult to find appropriate templates connecting the fragments in a stereochemically and synthetically reasonable way. The alternative is to start with a seed fragment in a certain region of the binding site, and then to append additional fragments in a step-wise build-up procedure (Fig. 1, center). An advantage of this approach is that chemical

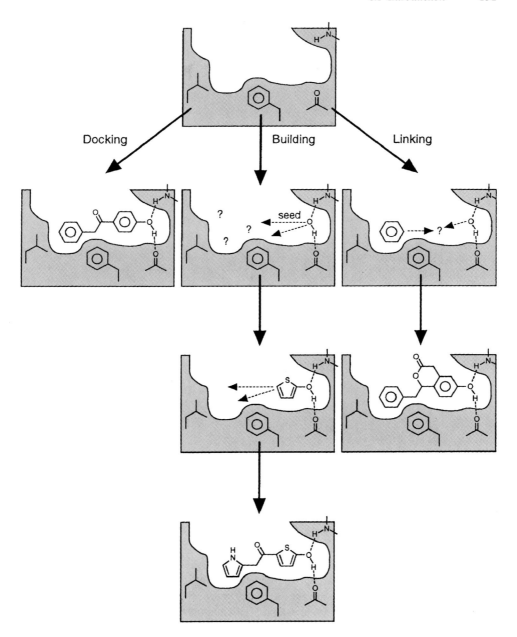

Figure 1. Different strategies for lead discovery by ligand docking. A complete ligand can be docked into the binding site (left). The *de novo* design of new ligands can be performed either by positioning a seed which is further extended by additional building blocks (center) or by simultaneous placement of several fragments that are then connected by suitable linking groups (right, sketch adapted from Verlinde and Hol [35]).

knowledge can be easily incorporated into the linking step. Therefore, synthetically accessible structures are more likely to be obtained by this approach. However, it tends to generate more flexible structures. As pointed out by Rotstein and Murcko [41], the build-up procedure can run into difficulties if a large gap between two separated regions of the binding pocket has to be bridged without the possibility to form extensive specific interactions with the protein in the gap region.

8.2   The Computer Program LUDI

8.2.1   Basic Methodology

LUDI is a fragment-based *de novo* design program which can be used both for 3D database searching and for the automatic construction of novel ligands either through building (step-by-step build-up) or through linking (placement of individual fragment and subsequent connection) [34, 55, 56]. As other programs for ligand design, LUDI requires basically three pieces of information: 1) the 3D structure of the target protein; 2) 3D structures of putative ligands or fragments for docking or constructing novel molecules; and 3) information about possible favorable interactions between the protein and the ligand.

In addition to the use of 3D protein structures as target structures, LUDI uses information derived from experimentally determined crystal structures of small organic molecules as stored in the Cambridge Structural Database (CSD) [57]. The X-ray structures are used in a two-fold way. First, a statistical analysis on preferred nonbonded contacts is used to define rules about favorable contact geometries in protein–ligand complexes. Second, the 3D structures of the small organic compounds themselves (or rigid substructures of them) are finally used as fragments for the docking into the binding site. In addition, the measured binding constants K_i of a large set of diverse protein–ligand complexes serve to calibrate the scoring function to prioritize the LUDI-hits. Finally, the 3D structure of known ligands can also be used to design automatically derivatives of the ligand to improve their binding characteristics.

An important conceptual aspect of LUDI is its ability to tolerate small uncertainties in the experimentally determined protein geometry. This uncertainty can give rise to unreasonably large contributions to the protein–ligand interaction energy if calculated with a molecular mechanics force field. In other words, interaction energies obtained from force field calculations can only be interpreted if a full geometry optimization including all degrees of freedom is carried out. We have decided to refrain from using a molecular-mechanics-based scoring. Instead, the error-tolerant empirical scoring function described in section 8.3 has been implemented in LUDI [58].

LUDI constructs novel protein ligands by joining molecular fragments. The program positions molecules or new substituents for a given lead into clefts of protein structures (e.g., an active site of an enzyme) in such a way that hydrogen bonds can be formed with the protein and hydrophobic pockets are filled with lipophilic groups. The positioning of fragments with LUDI is completely based on geometric operations and does not involve any force field calculations. The binding energy between the protein and a putative ligand is estimated after the docking procedure using a simple empirical scoring function that takes into account both enthalpic and entropic contributions. LUDI is a fully deterministic program with no use of random number generators. A LUDI calculation comprises the following steps:

(1) Generation of the interaction sites (positions suitable to form favorable interactions with the protein).
(2) Fit of fragments onto the interaction sites; two different modes of operation are possible: standard mode: unconstrained positioning in the protein binding site, link-mode: fragments are attached onto already positioned fragments.
(3) Scoring of the generated structures.

In a first step the program calculates 'interaction sites', which are discrete positions in space suitable to form hydrogen bonds or to fill a hydrophobic pocket. The interaction sites are derived from a statistical analysis of nonbonded contacts [56, 59] found in the Cambridge Structural Database (CSD) [57, 60, 61]. For every functional group of the protein, there exists not only a single position but a region in space suitable to form favorable interactions with the protein. In LUDI this full distribution of possible contact patterns is taken into account by using an ensemble of interaction sites distributed over the whole region of possible contact patterns. Fig. 2 visualizes the concept of interaction sites. The approach is purely geometrical and avoids costly calculations of potential functions. LUDI distinguishes between four different types of interaction sites: H-donor, H-acceptor, lipophilic-aliphatic and lipophilic-aromatic. In LUDI the H-donor and H-acceptor interaction sites are described by vectors (atom pairs) to account for the strong directionality of hydrogen bonds. The rules are described in detail elsewhere [56].

The second step is the fit of molecular fragments onto the interaction sites. The list of generated interaction sites is searched for suitable pairs, triplets, quadruples or pentuples to match the molecular fragments. The fit of the fragments is done by a root-mean-square (RMS) superposition onto the interaction sites. In the standard-mode, fragments are fitted independently from each other into the protein binding site. The fragments are taken from a library. We have generated several fragment libraries which are summarized in Table 1. The

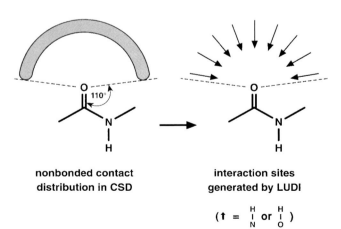

nonbonded contact distribution in CSD

interaction sites generated by LUDI

Figure 2. The concept of interaction sites in LUDI. The full range of possible hydrogen bond geometries as evidenced by a statistical analysis of nonbonded contact geometries in crystal packings of organic molecules (shown on the left side) is described by a set of interaction sites (shown as vectors on the right side).

Table 1. Fragment libraries for LUDI.

Type of fragments	No. of entries	Source of 3D coordinates
Small organic molecules ('Standard library')	1 100	Computer graphics + force field calculation
Small rigid molecules from the FCD (≤ 40 atoms, ≤ 2 rotatable bonds)	30 000	CONCORD
Small rigid molecules from the ACD (≤ 40 atoms, ≤ 2 rotatable bonds)	54 000	CORINA
Rigid substructures from the ACD	25 000	CORINA
Small organic molecules from the CSD	28 000	X-ray
Rigid substructures from the CSD	10 000	X-ray
Natural amino acids	20	X-ray (~ 100 conformers per amino acid)
Non-natural amino acids	50	Computer graphics + force field calculation

'standard'-library currently contains 1,100 diverse small molecules and was generated by us manually using computer graphics. The flexibility of some of the fragments in this library is accounted for by storing multiple conformers. LUDI can also be used to search larger fragment libraries. For example, we use structures from the 'Available Chemicals Directory' (ACD) and the Cambridge Structural Database (CSD) as fragments for LUDI [32]. In addition, we use rigid substructures from both databases as fragments. For large databases, storage of multiple conformers requires a significant amount of disk space. Therefore, in the current version, LUDI can treat fragment flexibility by generating several conformers 'on the fly'. The program automatically detects rotatable bonds, puts them into one of several categories (e.g., X-CH$_2$CH$_2$-X with allowed dihedrals 180°, 60°, −60°) and then generates several conformers.

LUDI can also be run in the 'link-mode'. In this mode, LUDI connects some or all of the fitted fragments by bridge fragments to form a single molecule. Alternatively, LUDI can append new fragments onto an already positioned fragment or lead compound. Several libraries are available for fragment linking. A comparatively small library of 1200 fragments was generated similarly to the 'standard'-library. In addition, libraries consisting of natural and non-natural amino acids have been prepared for the design of peptides. Furthermore, a fragment library generated by extracting rigid substructures from the ACD and the CSD can be used for linking. The concept of LUDI is summarized in Fig. 3.

When using LUDI, a possible strategy for *de novo* design is to carry out first a simple 3D search (running LUDI in standard-mode) using a 3D database and select a small number of diverse top scoring hits for biological testing. If experimentally satisfactory binding is observed for some of these structures then they are subsequently submitted to a further LUDI calculation in the link mode searching for substituents. Alternatively, if the 3D structure of protein complexed with a suitable lead is known, one can use this information to run LUDI in the link mode to search for derivatives.

The final step is the scoring of the generated protein–ligand complex. In LUDI, the problem of prioritization of the hits is approached by a two-step procedure. First, in a selection step, a number of criteria are applied in order to remove all structures with problems. Our

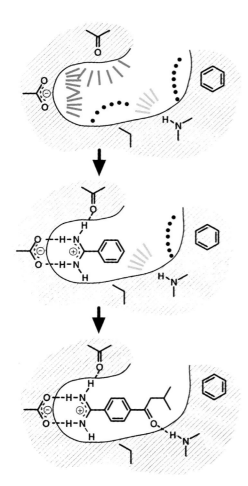

Figure 3. Basic steps of a calculation with LUDI. First, the interaction sites are generated (donor and acceptor interaction sites are depicted as bars, lipophilic interaction sites are depicted as circles). The next step is the fit of fragments onto the interaction sites. Finally, LUDI appends further fragments onto an already positioned fragment or lead compound

current strategy in selecting structures as possible ligands is to use a large number of different criteria and to allow for significant tolerances for each parameter. Currently, the following criteria are used:

(1) Only those fragments with a RMS deviation of the fit of the fragment onto the interaction sites below a certain threshold (typically 0.4–0.7 Å) are accepted.

(2) A further requirement for a successfully positioned fragment is that it does not overlap with the protein. However, small overlaps of up to 0.5 Å are tolerated, in order to account for induced fit effects.

(3) LUDI also checks for electrostatic repulsion between protein and ligand: if a polar ligand atom is closer to a protein atom of same polarity than a threshold distance (typically 3.5 Å for O..O contacts), then this particular positioning of the fragment is rejected. In the electrostatic repulsion check, only those protein atoms are taken into account that do not hydrogen bond with the ligand.

(4) The ligand should show good steric complementarity with the binding site. This implies that a large percentage of the ligand surface gets into contact with the protein surface upon binding. The percentage of the ligand surface that is buried when docked into the binding site is calculated and can also be used as a selection criterion.
(5) Good steric complementarity also implies that there are no cavities along the protein–ligand interface. LUDI can detect and measure cavities along the protein–ligand interface. It is possible to specify a upper limit for the cavity volume in the selection step.

In the prioritization step, all remaining structures that have not been rejected in the selection step are scored using the function described in section 8.3. It is possible to specify a lower limit for the score. LUDI will then accept only those structures with a score better than the user-defined threshold value. In addition, LUDI also tries to estimate the possible maximum score for each fragment (assuming a fully buried surface and the formation of hydrogen bonds with all polar groups of the fragment). The ratio actual score/possible maximum score can also be used as a selection criterion.

8.2.2 Applications of LUDI

In a validation study, we successfully applied LUDI to the design of inhibitors of dihydrofolate reductase and HIV-protease [56]. Pisabarro et al. [62] used a combination of GRID and LUDI to successfully design novel inhibitors of human synovial fluid phospholipase A_2 with enhanced activity. A calculation with GRID pointed to a lipophilic binding pocket not occupied by the lead compound. LUDI was then used to search for suitable substituents to fill this pocket. One suggestion from LUDI was synthesized and found to yield a ten-fold improvement in binding affinity.

In another application, a subset of $\sim 30\,000$ small molecules (with less than 40 atoms and 0–2 rotatable bonds) from the fine chemicals directory (FCD [63]) has been used in the search for possible novel ligands for four different proteins. The 3D structures were generated using the program CONCORD [64] (More recently, we also employed the program CORINA [65] for the 3D structure generation). For example, LUDI was applied to the search for ligands for the specificity pocket of the enzyme trypsin. The coordinates of the enzyme complexed with the inhibitor benzamidine [66] were used. The calculation using the 30 000 compound library takes 118 minutes on a Silicon Graphics Indigo R4000 workstation and retrieves 153 compounds as potential ligands for trypsin. LUDI calculates the highest score for *p*-methyl-benzamidine (**1**). The second highest score is found for benzamidine. Benzamidine binds trypsin with $K_i = 18$ μM [67]. *p*-Methyl-benzamidine was indeed found experimentally to bind trypsin with a slightly higher binding affinity than benzamidine [68]. A second example is the search for ligands of streptavidin. For this protein, the top scoring ligand retrieved by LUDI is 4-methyl-2-imidazolidinone (**2**). This was also found experimentally to bind streptavidin and avidin in the micromolar range [69]. The unsubstituted 2-imidazolidinone is also retrieved by LUDI. It is ranked as hit #9 still exibiting a very high score corresponding to K_i (predicted) < 10 μM in agreement with available experimental data [69].

Recently, Babine et al. used LUDI to design novel ligands for the FKBP-12 [70]. Starting from the known 3D structure of the protein, compound **3** was designed in a step-wise fash-

1   **2**   **3**   **Scheme 1**

ion. The compound **3** was synthesized and found to be a strongly binding ligand of FKBP-12 ($K_i = 12 \ \mu M$).

One of the strengths of LUDI is its ability to find small polar ligands for tight polar binding sites as present for example in trypsin. Our current experience indicates that for such proteins the program can retrieve interesting small molecules forming multiple hydrogen bonds with the protein. LUDI is now widely used in pharmaceutical industry [71].

8.3 Computational Methods to Predict Ligand Binding Affinities

Several computer programs are now available to create novel ligand structures that are complementary to a given binding site. These algorithms have the potential to create thousands of putative ligand structures overnight. The big problem is then to select some structures for synthesis. Therefore, the most important current challenge both in *de novo* design and 3D database searching is the development of new methods to prioritize the large number of diverse putative protein ligands resulting from such calculations. There is an urgent need for a fast and reliable ranking of the candidate structures.

The free binding enthalpy ΔG of a protein-ligand complex has an enthalpic and an entropic component ($\Delta G = \Delta H - T \Delta S$). The available experimental data [72–77] all show that both ΔH and ΔS contribute significantly to ligand binding. Moreover, in general there is no correlation between the binding enthalpy ΔH and the free binding enthalpy, ΔG.

This has important consequences for the theoretical description of protein–ligand interactions. A popular approach to investigate the energetics of protein–ligand interactions is to perform a molecular mechanics force field optimization and then to analyze the interaction energies of the optimized structure. Several groups have tried to correlate experimentally determined binding affinities with protein–ligand interaction energies obtained from a force field calculation. For one set of HIV-protease inhibitors [78] and for an ensemble of thrombin inhibitors [79], a good correlation was observed. However, this approach is unlikely to be generally applicable for several reasons. First, a pure force field approach does not account for desolvation effects and entropic effects. However, available experimental data clearly show, that ΔS is important and cannot be ignored. Large contributions to the calculated interaction energy arise from hydrogen bonds and ionic interactions. On the other hand, lipophilic contacts contribute little to the interaction energy as calculated by a force field. This approach is therefore in trouble, if the protein–ligand interaction is dominated by lipophilic interactions.

In principle, relative free enthalpies of binding can be obtained computationally, e.g., from free energy perturbation (FEP) or thermodynamic integration (TI) methods [80]. Some impressive results have been obtained using these techniques [81–86]. The FEP and TI calculations have provided valuable insights into the physical nature of specific protein–ligand interactions. For example, the relative binding affinity in a series of thermolysin inhibitors [85, 86] and HIV-protease inhibitors [82] were predicted correctly. The advantage of these methods is the rigorous treatment of basically all effects in protein–ligand interactions including solvent effects and conformational changes. In practice however, the methods suffers from convergence problems and therefore require very long simulation times. At best, they can be used for 10–100 ligand structures. In addition, at present the approaches are only capable to predict differences in binding affinities between closely related ligands. Therefore, it appears quite difficult to predict accurately the binding energies of protein–ligand complexes, even if the 3D structure of the complex is known. The rigorous modeling of such interactions is still a considerable challenge [87]. In view of the current limitations with the FEP and the TI approach, a number of alternative approaches have been evaluated to predict binding affinities. Recently, Warshel et al. [88] have shown that the semimacroscopic version of the Protein Dipoles Langevin Dipoles method (PDLD/S) can provide a good estimate for ΔG in a reasonable amount of computer time on current workstations.

In view of the importance of interactions with the solvent, a substantial amount of work has been invested into the development of methods calculating solvation energies, in particular free energies of solvation [89]. Empirical [90, 91] and quantum mechanical methods [92, 93] have been used to calculate solvation energies of small organic molecules.

A possible approach to assess the binding affinity of a large ensemble of ligands is to use a two-step procedure. First, a fast and simple empirical scoring function is used to prioritize the hits. Then, the top scoring structures are analyzed more rigorously using more accurate but computationally demanding approaches. Several simple scoring functions have been proposed [38, 94–97]. One simple scoring function will described below.

8.3.1 A New Scoring Function for Protein–Ligand Complexes

We have tested a simple empirical scoring function in the prediction of binding affinities for a protein–ligand complex of known 3D structure [38]. The purpose of the scoring function is to prioritize the hits obtained from computer program for *de novo* design or 3D database searching. The non-bonded protein–ligand interactions are assumed to be additive. The function takes into account hydrogen bonds, ionic interactions, the lipophilic protein–ligand contact surface and the number of rotatable bonds in the ligand. The following analytical form is used:

$$\Delta G_{binding} = \Delta G_0 + \Delta G_{hb} \sum_{h\text{-bonds}} f(\Delta R, \Delta \alpha)$$
$$+ \Delta G_{ionic} \sum_{ionic\ int.} f(\Delta R, \Delta \alpha)$$
$$+ \Delta G_{lipo} |A_{lipo}|$$
$$+ \Delta G_{rot} NROT$$

where $f(\Delta R, \Delta \alpha)$ is a penalty function which accounts for large deviations of the hydrogen bond geometry from ideality [38]. It tolerates small deviations of up to 0.2 Å and 30° from the ideal geometry that are often due to positional uncertainties in the X-ray structures.

ΔG_0 is a contribution to the binding energy that does not directly depend on any specific interactions with the protein but is associated with the overall loss of translational and rotational entropy of the ligand. ΔG_{hb} describes the contribution from an ideal hydrogen bond. ΔG_{ionic} represents the contribution from an unperturbed ionic interaction. ΔG_{lipo} represents the contribution from lipophilic interactions, assumed proportional to the lipophilic contact surface A_{lipo} between the protein and the ligand. The final parameter ΔG_{rot} describes the loss of binding energy due to the immobilization of internal degrees of freedom in the ligand. NROT is the number of rotatable bonds.

The data set used for the calibration of the function consisted of 45 protein–ligand complexes. For this set, the energy function reproduced the binding constants (ranging from 2.5×10^{-2} M to 4×10^{-14} M, corresponding to binding energies between -9 and -76 kJ mol^{-1}) with a standard deviation of 7.9 kJ mol^{-1} corresponding to 1.4 orders of magnitude in binding affinity. The individual contributions to protein–ligand binding obtained from the scoring function are: for an unperturbed uncharged hydrogen bond (ΔG_{hb}): -4.7 kJ mol^{-1}, an ideal ionic interaction (ΔG_{ionic}): -8.3 kJ mol^{-1}, lipophilic contact (ΔG_{lipo}): -0.17 kJ mol^{-1} Å^2, one rotatable bond in the ligand (ΔG_{rot}): $+1.4$ kJ mol^{-1}. The constant contribution (ΔG_0) obtained from the calibration is $+5.4$ kJ mol^{-1} (adverse to binding). The present approach is fast. For a given protein it allows the scoring of 10 small ligands per second on a current single-processor UNIX workstation.

We are currently investigating an improved scoring function. This function contains three additional new terms. The first new term (ΔG_{aro}) accounts for specific interactions between aromatic rings. The replacement of water molecules that form hydrogen bonds with the protein ($\Delta G_{H\text{-bonded water}}$) and the replacement of those water molecules that do not form a hydrogen bond ($\Delta G_{not\ H\text{-bonded water}}$) are considered as different contributions. In addition, the terms ΔG_{hb} and ΔG_{ionic} are now modulated by a new factor to discriminate between buried interactions and solvent accessible interactions. The new function is currently based on 80 protein–ligand complexes. The new scoring function is applicable to a broader range of protein–ligand complexes.

8.4 Current Challenges in Computational *de novo* Ligand Design

All current methods for *de novo* ligand design, including LUDI, face a number of limitations. Most importantly, the programs consider only interactions with the target protein. Transport properties, metabolic stability, or toxicity are not taken into account. It should also be noted that the current programs do not yet address the problem of the synthetic accessibility of the suggested structures. Further, it is clear that current methods for the prediction of the binding affinity need to be improved.

Another important limitation is that the protein is at present treated as rigid. A number of pairs of 3D protein structures with and without bound ligand have been determined and give some indication as to which conformational changes of the protein can happen during the binding of a ligand. Some proteins such as trypsin [65] or thrombin [98] have fairly rigid binding sites and do not exhibit large conformational changes upon ligand binding. Sometimes, a part of the protein (e.g., a loop) moves as a consequence of the ligand binding [99–101]. However, it has been shown that this movement is very similar for different ligands. Therefore, even for flexible proteins the 3D structure of a protein–ligand complex is a good starting point for *de novo* design programs.

8.5 Summary

The present review summarizes the current status in the development of new computer programs for the *de novo* design of protein ligands. As an example, LUDI has been described in some detail. This program is fast with typical execution times ranging from 20–300 seconds (using the standard library with 1100 fragments) and can therefore be used interactively. It should also be noted that LUDI is not limited to the design of protein ligands. It accepts any organic molecule as target structure and should therefore be of interest also to the design of host–guest complexes. The current experience indicates that LUDI is a useful addition to the toolbox of the medicinal chemist. Current efforts focus on improving the ability to predict accurately the binding energies of putative ligands, to construct only structures that are synthetically feasible, and to cope with the flexibility of both the ligand and the protein.

References

[1] Navia, M. A., and Murcko, M. A., *Curr. Opin. Struct. Biol.* **2**, 202–210 (1992)
[2] Baldwin, J. J., Ponticello, G. S., Anderson, P. S., Christy, M. E., Murcko, M. A., et al., *J. Med. Chem.* **32**, 2510–2513 (1989)
[3] Appelt, K., Bacquet, R. J., Bartlett, C. A., et al., *J. Med. Chem.* **34**, 1925–1934 (1991)
[4] Varney, M. D., Marzoni, G. P., Palmer, C. L., et al., *J. Med. Chem.* **35**, 663–676 (1992)
[5] Reich, S. H., Fuhry, M. A. M., Nguyen, D., et al., *J. Med. Chem.* **35**, 847–858 (1992)
[6] Thompson, W. J., Fitzgerald, P., Holloway, M. K., et al., *J. Med. Chem.* **35**, 1685–1701 (1992)
[7] Lam, P. Y. S., Jadhav, P. K., Eyermann, C. J., et al., *Science* **263**, 380–384 (1994)
[8] von Itzstein, M., Wu, W. Y., and Kok, G. B., *Nature* **363**, 418–423 (1993)
[9] Mack, H., Pfeiffer, Th., Hornberger, W., Böhm, H. J., and Höffken, H. W., *J. Enzyme Inhibition* **9**, 73–86 (1995)
[10] Cohen, N. C., Blaney, J. M., Humblet, C., Gund, P., and Barry, D. C., *J. Med. Chem.* **33**, 883–894 (1990)
[11] Martin, Y. C., *J. Med. Chem.* **35**, 2145–2154 (1992)
[12] Wireko, F. C., Kellogg, and G. E., Abraham, D. J., *J. Med. Chem.* **34**, 758–767 (1991)
[13] Kellog, G. E., Joshi, G. S., and Abraham, D. J., *Med. Chem. Res.* **1**, 444–453 (1992)
[14] Nicholls, A., Sharp, K. A., and Honig, B. H., *Proteins* **11**, 281 (1991)
[15] Connolly, M. L., *Biopolymers* **32**, 1215–1236 (1992)
[16] Lewis, R. A., J. *Comput. Aided Molec. Des.* **3**, 133–147 (1989)
[17] Kleywegt, G. J., and Jones, T. A., *Acta Crystallogr.* **D50**, 178–185 (1994)
[18] Goodford, P. J., *J. Med. Chem.* **28**, 849–857 (1985)
[19] Boobbyer, D. N. A., Goodford, P. J., McWhinnie, P. M., and Wade, R. C., *J. Med. Chem.* **32**, 1083–1094 (1989)
[20] Tomioka, N., and Itai, A., *J. Comput. Aided Molec. Des.* **8**, 347–366 (1994)
[21] Miranker, A., and Karplus, M., *Proteins: Struct. Funct. Genet.* **11**, 29–34 (1991)
[22] Blaney, J. M., and Dixon, J. S., *Perspect. Drug Disc. Design* **1**, 301–319 (1993)
[23] Kuntz, I. D., Meng, E. C., and Shoichet, B. K., *Acc. Chem. Res.* **27**, 117–123 (1994)

[24] DesJarlais, R. L., Sheridan, R. P., Seibel, G. L., Dixon, J. S., Kuntz, I. D., and Venkataraghavan, R., *J. Med. Chem.* **31**, 722–729 (1988)
[25] Meng, E. C., Shoichet, B. K., and Kuntz, I. D., *J. Comp. Chem.* **13**, 505–524 (1992)
[26] Meng, E. C., Gschwend, D. A., Blaney, J. M., and Kuntz, I. D., *Proteins* **17**, 266–278 (1993)
[27] Kuntz, I. D., *Science* **257**, 1078–1082 (1992)
[28] Shoichet, B. K., Stroud, R. M., Santi, D. V., Kuntz, I. D., and Perry, K. M., *Science* **259**, 1445–1450 (1993)
[29] Li, Z., Chen, X., Davidson, E., et al., *Chemistry & Biology* **1**, 31–37 (1994)
[30] Lawrence, M. C., and Davis, P. C., *Proteins* **12**, 31–41 (1992)
[31] Miller, M. D., Kearsley, S. K., Underwood, D. J., and Sheridan, R. P., *J. Comput. Aided Molec. Des.* **8**, 153–174 (1994)
[32] Böhm, H. J., *J. Comput. Aided Molec. Des.* **8**, 623–632 (1994)
[33] Lewis, R. A., and Leach, A. R., *J. Comput. Aided Molec. Des.* **8**, 467–475 (1994)
[34] Böhm, H. J., In: Kubinyi, H., Ed. 3D-QSAR in *Drug Design*, Escom Science Publishers; Leiden; 386–405 (1993)
[35] Verlinde, C. L. M. J., and Hol, W. G. J., *Structure* **2**, 577–587 (1994)
[36] (a) Nishibata, Y., and Itai, A., *Tetrahedron* **47**, 8985–8990 (1991); (b) Nishibata, Y., Itai, A., *J. Med. Chem.* **36**, 2921–2928 (1993)
[37] Bohacek, R. S., and McMartin, C., *J. Am. Chem. Soc.* **116**, 5560–5571 (1994)
[38] Gehlhaar, D. K., Moerder, K. E., Zichi, D., Sherman, C. J., Ogden, R. C., and Freer, S. T., *J. Med. Chem.* **38**, 466–472 (1995)
[39] Moon, J. B., and Howe, W. J., *Proteins* **11**, 314–328 (1991)
[40] Tschinke, V., and Cohen, N. C., *J. Med. Chem.* **36**, 3863–3870 (1993)
[41] Rotstein, S. H., and Murcko, M. A., *J. Med. Chem.* **36**, 1700–1710 (1993)
[42] Eisen, M. B., Wiley, D. C., Karplus, M., and Hubbard, R. E., *Proteins* **19**, 199–221 (1994)
[43] Caflish, A., Miranker, A., and Karplus, M., *J. Med. Chem.* **36**, 2142–2167 (1993)
[44] Lewis, R. A., *J. Mol. Graphics* **10**, 66 (1992)
[45] Gillet, V. J., Johnson, A. P., Mata, P., and Sike, S., *Tetrahedron Comput. Methods* **3**, 681–696 (1990)
[46] Gillet, V., Johnson, P., Mata, P., Sike, S., and Williams, P., *J. Comput. Aided Molec. Design* **7**, 127–153 (1993)
[47] Mata, P., Gillet, V. J., Johnson, P., Lampreia, J., Myatt, G. J., Sike, S., and Stebbings, A. L., *J. Chem. Inf. Comput. Sci.* **35**, 479–493 (1995)
[48] Rotstein, S. H., and Murcko, M. A., *J. Comput. Aided Molec. Des.* **7**, 23–43 (1993)
[49] Bartlett, P. A., Shea, G. T., Telfer, S. J., and Waterman, S. In: *Molecular Recognition: Chemical and Biological Problems,* Roberts S. M. (Ed.), Royal Society of London; 182–196 (1989)
[50] Pearlman, D. A., and Murcko, M. A., *J. Comput. Chem.* **14**, 1184–1193 (1993)
[51] Lewis, R. A., and Dean, P. M., *Proc. R. Soc. Lond.* **B236**, 125–140 (1989)
[52] Lewis, R. A., and Dean, P. M., *Proc. R. Soc. Lond.* **B236**, 141–162 (1989)
[53] Chau, P. L., and Dean, P. M., *J. Comput. Aided Molec. Des.* **6**, 385–396 (1992)
[54] VanDrie, J., Weininger, D., and Martin, Y. C., *J. Comput. Aided Molec. Des.* **3**, 225–251 (1989)
[55] Böhm, H. J., *J. Comput. Aided Molec. Des.* **6**, 61–78 (1992)
[56] Böhm, H. J., *J. Comput. Aided Molec. Des.* **6**, 593–606 (1992)
[57] Allen, F. H., Bellard, S., Brice, M. D., Cartwright, B. A., Doubleday, A., Higgs, H., Hummelink-Peters, T., Kennard, O. Motherwell, W. D. S., Rodgers, J. R., and Watson, D. G., *Acta Crystallogr.* **B35**, 2331–2339 (1979)
[58] Böhm, H. J., *J. Comput. Aided Molec. Des.* **8**, 243–256 (1994)
[59] Klebe, G., *J. Mol. Biol.* **237**, 212–235 (1994)
[60] Allen, F. H., Kennard, O., and Taylor, R., *Acc. Chem. Res.* **16**, 146–153 (1983)
[61] Allen, F. H., Davies, J. E., Galloy, J. J., et al., *J. Chem. Inf. Comput. Sci.* **31**, 187–204 (1991)
[62] Pisabarro, M. T., Ortiz, A. R., Palomar, A., Cabre, F., Garcia, L., Wade, R. C., Gago, F., Mauleon, D., and Carganico, G., *J. Med. Chem.* **37**, 337–341 (1994)
[63] The fine chemicals directory (FCD) and the available chemicals directory (ACD) are distributed by Molecular Design Ltd., San Leandro, 2132 Farallon Drive, CA 94577
[64] Program CONCORD, distributed by Tripos Ass., 1699 S. Hanley Rd., St. Louis, MO 63144
[65] Sadowski, J., Rudolph, C., and Gasteiger, J., *Tetrahedron Comput. Methodol.* **3**, 537 (1990)
[66] Marquart, M., Walter, J., Deisenhofer, J., Bode, W., and Huber, R., *Acta Crystallogr.* **B39**, 480–490 (1983)
[67] Mares-Guia, M., and Shaw, E., *J. Biol. Chem.* **240**, 1579–1585 (1965)
[68] Recanatini, M., Klein, T., Yang, C. Z., McClarin, J., Langridge, R., and Hansch, C., *Mol. Pharmacol.* **29**, 436–446 (1986)

[69] Green, N. M., *Adv. Protein Chem.* **29**, 85 (1975)
[70] Babine, R. E., Bleckman, T. M., Kissinger, C. R., Showalter, R., Pelletier, L. A., Lewis, C., Tucker, K., Moomaw, E., Parge, H. E., and Villafranca, J. E., *Biorg. Med. Chem. Lett.* **15**, 1719–1724 (1995)
[71] LUDI is available from MSI Technologies, 9685 Scranton Road, San Diego, CA 92121–2
[72] Wieseman T., Wiliston S., Brandts J., and Lin L., *Anal. Biochem.* **179**, 131 (1989)
[73] Hitzemann, R., *Trends Pharmacol. Sci.* **9**, 408–411 (1988)
[74] Hitzemann, R., Murphy, M., and Currell, J., *Eur. J. Pharmacol.* **108**, 171–177 (1985)
[75] Weiland, G. A., Minnemann, G. P., and Molinoff, P. B., *Mol. Pharmacol.* **18**, 341–347 (1980)
[76] Epps, D. E., Cheney, J. Schostarez, H., Sawyer, T. K., Prairie, M., Krueger, W. C., and Mandel, F., *J. Med. Chem.* **33**, 2080 (1990)
[77] Weber, P. C., Wendoloski, J. J., Panoliano, M. W., and Salemme, F. R., *J. Am. Chem. Soc.* **114**, 3197–3200 (1992)
[78] Holloway, M. K., Wai, J. M., Halgren, T. A., et al., *J. Med. Chem.* **38**, 305–317 (1995)
[79] Grootenhuis, P. D. J., and Van Galen, P. J. M., *Acta Crystallogr.* **D51**, 560–566 (1995)
[80] van Gunsteren, W. F., and Weiner, P. K., *Computer Simulations of Biomolecular Systems,* Escom; Leiden, 1989
[81] Reddy, M. R., Viswanadhan, V. N., and Weinstein, J. N., *Proc. Natl Acad. Sci. USA* **88**, 10287–10291 (1991)
[82] Rao, B. G., Tilton, R. F., and Singh, U. C., *J. Am. Chem. Soc.* **114**, 4447–4452 (1992)
[83] Miyamoto, S., and Kollman, P. A., *Proteins* **16**, 226–245 (1993)
[84] Miyamoto, S., and Kollman, P. A., *Proc. Natl Acad. Sci. USA* **90**, 8402–8406 (1993)
[85] Bash, P. A., Singh, U. C., Brown,F. K., Langridge, R., and Kollman, P. A., *Science* **235**, 574–576, (1987)
[86] Merz, K. M., and Kollman, P. A., *J. Am. Chem. Soc.* **111**, 5649–5658 (1989)
[87] Kollman, P. A., *Curr. Opin. Struct. Biol.* **4**, 240–245 (1994)
[88] Warshel, A., Tao, H., Fothergill, M., and Chu, Z. T., *Isr. J. Chem.* **34**, 263–256 (1994)
[89] Honig, B., and Nicholls, A., *Science* **268**, 1144–1149 (1995)
[90] Kang, Y. K., Gibson, K. D., Nemethy, G., and Scheraga, H. A., *J. Phys. Chem.* **92**, 4739–4742 (1988)
[91] Williams, R. L., Vila, J., Perrot, G., and Scheraga, H. A., *Proteins* **14**, 110–119 (1992)
[92] Cramer, C. J., and Truhlar, D. G., *J. Am. Chem. Soc.* **113**, 8305 (1991)
[93] Cramer, C. J., and Truhlar, D. G., *J. Comp. Aided Molec. Des.* **6**, 629–666 (1992)
[94] Andrews, P. R., Craik, D. J., and Martin, J. L., *J. Med. Chem.* **27**, 1648–1657 (1984)
[95] Vajda, S., Weng, Z., Rosenfeld, R., and DeLisi, C., *Biochemistry* **33**, 13 977–13 988 (1994)
[96] Krystek, S., Stouch, T., and Novotny, J., *J. Mol. Biol.* **234**, 661–679 (1993)
[97] Horton, N., and Lewis, M., *Protein Sci.* **1**, 169–181 (1992)
[98] Banner, D. W., and Hadvary, P., *J. Biol. Chem.* **266**, 20085–20093 (1991)
[99] Sali, A., Veerapandian, B., Cooper, J. B., Moss, J. B., Hofmann, T., and Blundell, T. L., *Proteins* **12**, 158–170 (1992)
[100] Rahuel, J., Priestle, J. P., and Grütter, M. G., *J. Struct. Biol.* **107**, 227–236 (1991)
[101] Wierenga, R. K., Noble, M. E. M., and Davenport, R. C., *J. Mol. Biol.* **224**, 1115–1126 (1992)

9 The Future of Structure-Based Design: A Worthy Precept?

H. J. Böhm and K. Gubernator

9.1 Introduction

The present volume on structure-based ligand design could only highlight some of the many facets of this fascinating area of life science research. Nevertheless, the examples presented clearly demonstrate that the knowledge of the 3D structure of the target protein or close homologs can be successfully exploited to contribute to the discovery of new drugs with superior properties. The work described by Lunney and Humblet on renin and HIV protease inhibitors (Chapter 3), Borkakoti on zinc metalloproteases (Chapter 4), Gubernator et al. on beta-lactamase inhibitors (Chapter 5) and Taylor on sialidase-inhibitors (Chapter 6) has led to several interesting development candidates. Some of these are now under clinical investigation. Structure-based design clearly played a major role in the discovery of the carbonic anhydrase inhibitor, dorzolamide (Chapter 2). This drug has already become the most widely prescribed antiglaucoma product in the US within its first year on the market.

So far, structure-based ligand design has probably had its most significant impact on the development of HIV-protease inhibitors. The 3D structures of non-viral aspartic proteases had already paved the way towards the design of initial inhibitors which were further tailored for the viral enzyme using the information from the 3D structures of HIV-protease complexes. The 3D structures of the HIV-protease–inhibitor complexes have been determined for all compounds on the market (Fig. 1) or in advanced clinical trials. All techniques of structure-based design have been successfully employed in the design of novel inhibitors – and the story continues. Currently, second-generation compounds are designed which will hopefully prove useful to combat resistant strains of the virus. Again, structure-based design plays a vital role [1, 2]. The design of thrombin inhibitors is another area where the knowledge of the 3D structure has been successfully used in the design of selective high-affinity inhibitors [3–5]. Many more examples could be cited [6–13].

In our experience, structure-based design is most valuable if it is used in the very early stages of a project. The knowledge of the 3D structure from the beginning of a drug discovery effort can lead to completely new leads via computational tools such as docking and *de novo* design. It also allows to understand the binding mode of ligands discovered by screening and can thus guide the further lead optimization.

9.2 Development in the Design Process

Structure-based ligand design is a rapidly developing field. Significant methodological advances both in experimental structure determination and in computational ligand design

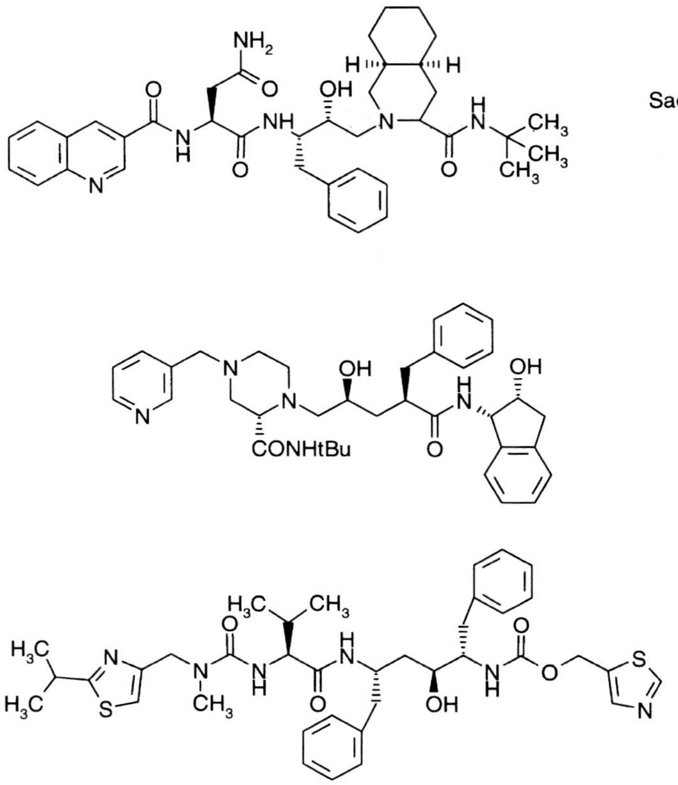

Figure 1. Structures of marketed HIV-protease inhibitors.

techniques can be expected in the next 5 years. For example, it was shown recently for the serine protease elastase that preferred binding sites of small organic molecules such as acetonitrile can be determined experimentally [14, 15]. If this technique can be further refined so that it can be applied on a regular basis to a broad range of proteins, it will have an enormous impact on the drug discovery process. In our opinion, the most important challenge in the further development of computational tools for ligand design is the improvement of existing methods for the prediction of binding affinities. If this can be combined with procedures that take into account the synthetic accessibility of the designed compound, then a true breakthrough will be achieved. At present, several groups are actively involved in the development of such second-generation tools for structure-based design. We believe, that this goal can be best achieved with an interdisciplinary effort, involving structural biologists, synthetic and computational chemists. Not surprisingly, a significant part of this basic research is presently done within the pharmaceutical industry.

The pharmaceutical industry is working in a rapidly changing environment. The question then arises: What role will structure-based design play in the future? First, structure-based ligand design will continue to benefit from the progress in structural molecular biology. As more and more groups start to work on the structure determination of biopolymers, an even larger number of solved 3D structures will be added to the data pool, allowing the number of projects where structure-based ligand design can be applied to further increase. We expect that this will also hold for therapeutic areas such as the CNS where the use of structure-based ligand design is not so widespread at present due to lack of structural data on molecular targets. Even if the structure determination of membrane-bound receptors may prove difficult in the near future, in some cases focussing just on the ligand-binding domain may prove to be fruitful. Second, there is the continuing challenge of understanding the mechanism of action of existing and future drug on a molecular basis. How does a compound bind to the target protein? What is the origin of its selectivity? These questions can only be answered unambigously if 3D structural information is available.

Nevertheless, there is also a growing pressure on pharmaceutical industry to improve the efficiency and to reduce expenses. For example, the pharmaceutical company GlaxoWellcome has announced that their goal for the year 2000 is to bring three significant new drugs to the market per year. (Before the merger of Glaxo and Wellcome, the combined rate of both companies was roughly one new major drug per year.) Other major pharmaceutical companies have similar goals. This goal can only be achieved if the process of drug discovery is expedited dramatically. Structure-based design will only continue to play a key role in industrial pharmaceutical research if it can contribute to this acceleration. We are confident that this will be indeed the case for a number of reasons.

9.3 Screening Systems

One of the recent developments in drug discovery is high throughput screening. Large pharmaceutical companies have compound collections consisting of 100 000–1 000 000 compounds. These substances are tested in biological assays and any compounds showing interesting activity are then investigated further. Significant efforts are underway at all major pharmaceutical companies to increase the size of their compound collections through substance aquisition and by in-house syntheses. Indeed, high throughput screening has proven to be particularly useful in the discovery of new antagonists for G-protein-coupled receptors. However, success rates appear to be much smaller for other classes of therapeutic targets such as for example serine proteases. Therefore, even with very large substance collections at hand, pharmaceutical companies sometimes face the problem that no useful hits emerge from the screening. In this case structure-based design remains the viable alternative for lead discovery. The current volume examplifies several cases where structure-based design was the key to the success of the drug discovery project.

Second, if we consider computational docking as a suitable option to complement experimental screening, the major advantage is the very large number of potential ligands that can be evaluated. For example, if we consider a simple chemical reaction such a the four-component Ugi-reaction [16], a reaction of an isonitrile R_1-NC, an aldehyde, R_2-CHO, an amine R_3-NH_2 and a carboxylic acid R_4-COOH, the use of 100 isonitriles, 1000 aldehydes, 1000 amines and 1000 carboxylic acids allows for the formation of 10^{11} Ugi-Products – at least in theory.

$$R_1{-}NC \ + \ R_2{-}CHO + R_3{-}NH_2 + R_4{-}COOH \longrightarrow$$

This number is far beyond the testing capabilities of any existing biological assay. However, by using advanced computational techniques such as combinatorial docking, the handling of this large dataset appears feasible. In this approach, the building blocks are positioned in the binding site individually and linked according to the desired chemical reaction. At present, computational docking cannot replace experimental testing because the current methods for affinity prediction are not accurate enough. Nevertheless, by taking into account the uncertainties of the computational approach, sets of compounds can be generated which have a high probability to bind to the target protein. In this way computational techniques can be used to direct the synthesis of biased libraries which are specifically designed to bind to selected targets.

In our opinion, one of the major benefits of computational structure-based design techniques such as docking and *de novo* design is their potential ability to discover ligands with low molecular weights. The knowledge of the 3D protein structure is optimally exploited if a designed molecule having perfect steric and electronic complementarity with the target binds to the protein in a low-energy conformation. The prudent use of all features of a protein binding pocket can lead to molecules of low molecular weight with high affinity. A molecule with a molecular weight below 300 Da and a K_i-value below 10 μM is much more amenable to optimization than a 'baroque' natural product with a high molecular weight, several stereo centers and nanomolar binding affinity. We expect that in the future the most significant impact of structure-based design will come from the design of such simple 'Bauhaus-Style' ligands.

9.4 Future Prospects

So far, most of the structure-based design has focused on the design of protease inhibitors, because structural information was predominantly available for this class of proteins. In the future, as more and more 3D structures of therapeutically relevant targets become available and their mechanim is understood, the range of applicability will broaden. For example, the recently published structure of the extracellular fragment of a T-cell receptor bound to a class I MHC-peptide complex [17] opens up a new route to the design of drugs that interact with this system. Furthermore, the stunning recent progress in the field of genomics is creating an enormous wealth of sequence data which will change the way that new targets are selected in the pharmaceutical industry. More and more information becomes available in several fields, for example on signal transduction. It is obvious, that kinases and phosphatases play an essential role in signal transduction, and one future challenge of structure-based ligand design will undoubtedly be to apply the arsenal of available tools to the design of selective kinase and phosphatase inhibitors, as well as to other emerging classes of therapeutically interesting targets.

In summary, structure-based ligand design will continue to play a central role in the drug discovery process. With significant methodological advances in sight, one can expect that the influence will be even more far reaching in the future.

Acknowledgement

We would like to thank Neera Borkakoti for helpful discussions.

References

[1] Anderson, P. S., Kenyon, G. L., and Marshall, G. R. (Eds.) Therapeutic approaches to HIV. *Persp. Drug Discov. Design* **1**, 1–128 (1993)

[2] Thairivongs, S., Skulnick, H. I., Turner, S. R., Strohbach, J. W., Tommasi, R. A., Johnson, P. D., Aristoff, P. A., Judge, T. M., Gammill, R. B., Morris, J. K., Romines, K. R., Chrusciel, R. A., Hinshaw, R. R., Chong, K. T., Tarpley, W. G., Poppe, S. M., Slade, D. E., Lynn, J. C., Horng, M. M., Tomich, P. K., Seest, E. P., Dolak, L. A., Howe, W. J., Howard, G. M., Schwende, F. J., Toth, L. N., Padbury, G. E., Wilson, G. J., Shiou, L., Zipp, G. L., Wilkinson, K. F., Rush, B. D., Ruwart, M. J., Koeplinger, K. A., Zhao, Z., Cole, S., Zaya, R. M., Kakuk, T. J., Janakiraman, M. N., and Watenpaugh, K. D., *J. Med. Chem.* **39**, 4349–4353 (1996)

[3] Obst, U., Gramlich, V., Diederich, F., Weber,L., and Banner, D. W., *Angew. Chem. Int. Ed. Engl.* **107**, 1874–1877 (1995)

[4] Hilpert, K., Ackermann, J., Banner, D. W., Gast, A., Gubernator, K., Hadvary, P., Labler, L., Müller, K., Schmid, G., Tschopp, T., and van de Waterbeemd, H., *J. Med. Chem.* **37**, 3889–3901 (1994)

[5] Mack, H., Pfeiffer, T., Hornberger, W., Böhm, H. J., and Höffken, H. W., *J. Enzyme Inhib.* **9**, 73–86 (1995)

[6] Montgomery, J. A., *Med. Res. Rev.* **13**, 209–228 (1993)

[7] Walkinshaw, M. D., *Med. Res. Rev.* **12**, 317–372 (1992)

[8] Navia, M. A., and Murcko, M. A., *Curr. Opin. Struct. Biol.* **2**, 202–210 (1992)

[9] Bedell, C. R. (Ed.) *The Design of Drugs to Macromolecular Targets.* Wiley, Chichester, 1992

[10] Greer, J., Erickson, J. W., Baldwin, J. J., and Varney, M. D., *J. Med. Chem.* **37**, 1035–1054 (1994)

[11] Verlinde, C. L. M. J., and Hol, W. G. J., *Structure* **2**, 577–587 (1994)

[12] Bohacek, R., McMartin, C., and Guida, W. C., *Med. Res. Rev.* **16**, 3–50 (1996)

[13] Böhm, H. J., and Klebe, G., *Angew. Chem. Ind. Ed. Engl.* **35**, 2588–2614 (1996)

[14] Allen, K. N., Bellamacina, C. R., Ding, X., Jeffery, C. J., Mattos, C., Petsko, G. A., and Ringe, D., *J. Chem. Phys.* **100**, 2605–2611 (1996)

[15] Mattos, C., and Ringe, D., *Nature Biotech.* **14**, 595–599 (1996)

[16] Weber, L., Wallbaum, S., Broger, C., and Gubernator, K., *Angew. Chem. Int. Ed. Engl.* **34**, 2280–2282 (1995)

[17] Garcia, K. C., Degano, M., Stanfield, R. L., Brunmark, A., Jackson, M. R., Peterson, P. A. Teyton, L., and Wilson, I. A., *Nature* **274**, 209–219 (1996)

Subject Index